ASHEVILLE-BUNCOMBE TECHNICAL INSTITUTE

NORTH CAROLINA
STATE BOARD OF EDUCATION
DEPT. OF COMMUNITY COLLEGES
LIBRARIES

DISCARDED

JUN 1 2 2025

Project Management Using Network Analysis

Other Titles in this Series
Harrison   Capital Investment Appraisal
Hobbs   Control Over Inventory and Production
Gorle and Long   Essentials of Product Planning
Consulting Editorship: Metra Consulting Group

Other McGraw-Hill Management Manuals
Davies   The Organization of Training
Margerison   Managing Effective Work Groups
Drake and Smith   Behavioural Science in Industry

Project Management Using Network Analysis

H. R. Hoare
Group Corporate Planner,
Manbré & Garton, Ltd., London

London · New York · St Louis · San Francisco · Düsseldorf · Johannesburg
Kuala Lumpur · Mexico · Montreal · New Delhi · Panama · Paris · São Paulo
Singapore · Sydney · Toronto

Published by
McGRAW-HILL Book Company (UK) Limited
MAIDENHEAD . BERKSHIRE . ENGLAND

07 084416 X

Copyright © 1973 McGraw-Hill Book Company (UK) Limited, except where otherwise acknowledged. All rights reserved. No part of this publication may be reproduced, stored in a retrieval system, or transmitted, in any form or by any means, electronic, mechanical, photocopying, recording, or otherwise, without the prior permission of McGraw-Hill Book Company (UK) Limited, or of the original copyright holder.

Printed and bound in Great Britain

Contents

	Preface	vii
Chapter		
1	Networks: Their uses and advantages	1
2	Identifying the problem	7
3	Preparing for the network	11
4	Network diagram	20
5	Time analysis of the network	36
6	Critical path and float	45
7	Extension of the analysis	57
8	Resource allocation	66
9	Project control	85
10	Network analysis within an organization	95
	Glossary	104
	Bibliography	107
	Index	108

To my wife, Gillian

Preface

Many books are available on the subject of network analysis and there are many examples of networks being used in practice. However, most of the available books are textbooks for actual practitioners and most of the examples provide great detail of the techniques. There seem to be no books available for directors and managers who are going to be affected by, but not directly involved in, projects using network analysis. Therefore, I hope that the time is right to write a book on networks for them. The book should also provide a first appreciation for those who are going to be technically involved in the use of network analysis.

Thus, the primary object of the book is to provide the reader with sufficient knowledge to be able to question networks and to be able to provide constructive criticism of the development of project plans and controls. So often, new techniques are installed without question and the results then do not aid managers. However, if a manager can be critical of the outputs, and can assist by knowing the limitations and advantages of the technique, then the many advantages of network analysis can be realized.

These advantages have led to the rapid acceptance of the techniques over the last fifteen years. In particular, network analysis provides means:

1. to plan projects so that the objectives, in terms of time and cost, can be evaluated;
2. to control projects so that as soon as the actual performance begins to differ from the plan, remedial action can be taken if necessary;
3. to provide a means of communication between the various departments and companies involved in a project;
4. to provide a discipline to the organization by setting down methods of working explicitly;
5. to improve the quality of estimation and implementation of projects.

In the book, I have tried to avoid jargon, although network analysis has already developed its own language (a Glossary of terms is included in the book for reference). The principles of network are in fact simple to describe, and there is no need for advanced mathematical knowledge to explain them. For the most part, I have tried to maintain a descriptive approach, rather than a numerical exposition.

I would like to acknowledge the help, in preparing this book, of Mr Hugh Sutherland and the many secretaries who have been involved in typing the manuscript. I acknowledge the permission of International Computers Ltd to use certain ICL material in this publication. Save for such permission, all copyright, patent, and other intellectual property-rights to ICL material belong to ICL.

H. R. Hoare

1
Networks: Their uses and advantages

Wherever the manager turns these days, he encounters references to *network analysis*. Yet it was only in 1957 that the first networks were developed for project planning and control and their results published. The two pioneering systems were independently conceived in the USA.

THE PIONEERING METHODS

PERT One of these methods of network analysis, developed by the Special Projects Office of the US Navy, is known as PERT, which is a mnemonic for Programme Evaluation and Review Technique. It was used to plan and control the design and development of the *Polaris*-missile project. In this first application, the technique is claimed to have saved some two years over the completion date, and it was immediately adopted by American Government Departments as a standard requirement in most of their large-scale projects.

This initial use of PERT coped with an extremely complex problem, full of uncertainty and open to many unknown influences; the several-thousand subcontractors involved in this large project had to be co-ordinated, and PERT provided an adequate means of communication. Control was based on the timing of the completion of certain phases within the projects, irrespective of the cost involved. Uncertainty in the duration of the tasks, due to the considerable amount of research and development necessary, was introduced to allow: (i) calculations of probabilities of achieving deadlines; and (ii) evaluation of the effect of action to ensure such achievement with a high degree of certainty.

CPM The other pioneering system of network analysis, developed at the same time, is known as CPM, which is the mnemonic for Critical Path Method. This technique was developed by the American chemical-company, Du Pont de Nemours.

The object was to control the work required for an overhaul of a large chemical-plant. Overall costs were important, and the plans were designed to minimize them. By reducing the time of the various critical tasks—i.e., tasks whose duration directly affected the total project duration—extra costs, in terms of resources of men, materials, and/or machinery, had to be incurred. Against this, the saving in down-time of the plant could be evaluated in financial terms and then compared with the increasing costs of shortening the critical tasks. Finally, a least-cost schedule could be obtained. Figure 1.1 shows how these relationships provide an optimum solution.

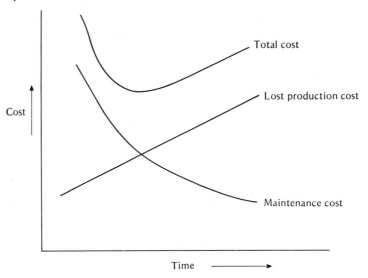

Fig. 1.1

The basic difference between PERT and CPM is in the objectives which their application is meant to attain. In using PERT, the US Navy was primarily interested in reducing the duration of the *Polaris* project, irrespective of cost. In using CPM, Du Pont were primarily concerned with cutting the overall costs of the plant overhaul.

FURTHER DEVELOPMENTS The use of networks has now expanded, not only in terms of the techniques of preparing a network, but also regarding the type of project to which network planning and control may be applied. Unfortunately, this has led to a proliferation of mnemonics, names, and approaches—all basically the same technique, with variations to provide greater sophistication. They include: PEP (Programme Evaluation Procedure); CPA (Critical Path Analysis); and computer trade-names, such as LESS (Least-Cost Estimating and Scheduling). Such developments were based mainly on the use of the network to minimize the time taken to complete projects and to evaluate the associated costs. A further development has been introduced to deal with resources requirements and constraints by which the planning of a project can be influenced. For example, an objective might be to minimize peak-demand for resources at certain stages. These problems have been solved by a method of approximation, as there is no automatic way of evaluating optimum use of resources.

A particular development of network diagramming was first used a few years after PERT and CPM. The method was developed simultaneously within Metra International in France, and known as the Metra Potential Method (MPM), the CEGB, in England (Precedence Diagrams) and in other countries. In this method, the diagram conventions are different from the conventions used in PERT and CPM.

ESSENTIALS AND ADVANTAGES

Networks are essentially a technique to aid management in the planning and control of projects. In general, the network analysis can be applied advantageously to those projects involving specific start-times and end-times. Networking is particularly useful where a large number of interrelated tasks are to be carried out, any of which may occur simultaneously.

A network diagram represents diagrammatically the tasks, or activities, that have to be carried out. In PERT, CPM, and techniques developed from them, tasks are shown as arrowed lines, each beginning and ending at an identifiable point of time. These points are called events, and are usually represented as circles on the diagram. The network also shows the relationships between the different tasks based on physical constraints or current practice. Figure 1.2 shows a typical section of a network diagram drawn with these conventions.

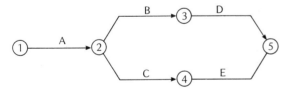

Fig. 1.2

However, it is possible to represent each activity within a box and to develop the logical sequence between the activities by arrows, representing constraints or practice. These were the conventions developed for MPM and Precedence Diagrams.

Structuring the Project

The first stage in preparing a network is to have clearly defined objectives for the project to be implemented. This is vital to the application of network analysis. Once the objectives have been set, then the network can be constructed, based on the logical relationships between the various tasks. Given these relationships, it is

3

possible to prepare a network diagram. Then, using this diagram, evaluation of the overall duration of the project is possible, together with the times by which certain critical tasks should be completed and the requirements for resources.

Once the results of this evaluation have been compared with the objectives, previously set, a decision can be taken on the necessary action required to bring the forecast results in line with the objectives. Finally, a plan of action can be specified, giving the times at which tasks should start and by which they should be completed, together with the resources that should be allocated. During the execution of the project, the actual achievement, against plan, can be fed back to the management. The effects of any deviations are then evaluated (using the network) and new plans of action are developed and issued.

Other advantages of applying network analysis are:

(a) It provides a means of communication between the various departments involved in a project and, at the same time, between the senior managers and those who are implementing it.
(b) It prescribes a discipline of thought to be applied before the project begins, so that actions are carefully evaluated from the outset.
(c) It permits careful appraisal of alternative methods of completing the project, concentrating on tasks which are critical to its achievement.
(d) It permits clear definitions of responsibility.
(e) It makes the collection of data and statistics a formal function within an organization.

Use of Bar Charts Bar charts, such as Gantt charts, have been used to plan and control projects, but networks have a number of comparative advantages over them. In particular, bar charts do not readily show tasks which are critical to the achievement of the overall project objectives, nor the degree of flexibility in the timing of those activities which are not critical. Moreover, it is not easy with a bar chart to evaluate the changes necessary in the plan of action, either when a decision is taken to modify tasks, or when a task does not achieve, in execution, the planned timetables.

Many planning tools, such as wall charts, have been designed to provide some flexibility in representing tasks; but with a large project, involving many tasks, it is almost impossible to remember all the subsequent tasks that will be affected by a change in the duration of any one. Therefore, with many activities occurring concurrently, each with a possibility of change in actual timings,

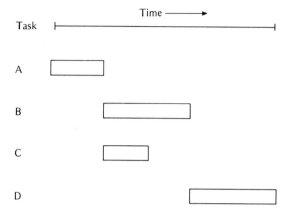

Fig. 1.3

the problems of control become most unwieldy. Figure 1.3 might be the bar chart of the tasks involved in Fig. 1.2. From this, it is not possible to know if a delay in activity B will have an effect on the timings of other activities.

However, bar charts do have their uses within a project planned and controlled by network. In particular, once activities have been evaluated in terms of their timings and the corresponding degrees of flexibility, such charts are useful in a manual assessment of alternative resource allocations. Of course, if a computer is used, there is no value in preparing a bar chart for this purpose.

One of the limitations of a network diagram is that it does not represent, visually, the time-scale involved; nor does it facilitate rapid indentification of a particular task. These two aspects are particular strengths of bar charts. Therefore, once the plan of action has been prepared, either by hand or as output from a computer, it may be useful to plot the schedule on a Gantt chart, for ready recognition of the relative timing of the activities.

Without doubt, network diagramming and analysis for the planning and controlling of projects provides a valuable and powerful tool to aid managers. Networks can be applied to small as well as large projects, and in many circumstances it is not necessary to use a computer. Projects involving more than, say, twenty defined tasks can benefit from the application of network analysis. Computers play their part in projects involving large numbers of tasks with complex interrelationships, particularly when, during implementation, the control evaluation of the results to date may have to be frequently made. Finally, the basic principles of network diagramming and analysis are extremely simple and involve no advanced mathematical techniques. This ensures very rapid understanding by

Power and Versatility

those managers not involved in the detailed development of a project network.

But network analysis is *not* a panacea: it does not solve all the problems of the project manager, and there is still a considerable area in which human decisions must be taken. For example, network analysis does not do the planning directly. It simply provides a tool that project management can use for planning, and it does so by supplying management with better information for making decisions on plans, schedules, and control.

Checklist

Network analysis started only fifteen years ago. The first applications were known as PERT and CPM. Are you quite clear about the distinction between them?

Since those early days, network analysis has been used more and more and has proved its value time and again. In particular, it can be applied to projects with specific start- and end-times, and it has many advantages over the conventional bar charts. Do you fully appreciate what these advantages are?

Are you using network analysis? If not, have you consciously made a decision *not* to apply it? If that is so, what are your main objections?

Consider whether there are any projects in your organization which could benefit from the use of this technique. If you see none at this stage, will you review the matter after reading a few more chapters?

2
Identifying the problem

The technique of network analysis has been most successful in planning and controlling projects with definite start- and endpoints. These points are fixed in time, either as objectives or as the result of a broad estimation of the tasks to be performed, and they provide an overall assessment of achievement in the process of implementation. Essentially, there are two conditions required for the application of network techniques: (i) the ability to specify sequential relationships between activities; and (ii) to provide estimated durations for activities (note that work content is not relevant—some judgement of the resources allocated is necessary). More specifically, there are some types of major undertaking for which networks are the most suitable means of planning and control.

AREAS OF APPLICATION

Two of the main areas of application are typified by the projects for which network analysis was first developed: large-scale maintenance or overhaul work; and research and development. Networks are also applied to projects of construction, and to those for introducing new products and processes. There are many miscellaneous applications as well, and the scope is increasing continuously; networks are currently being used in production control, and are being developed for use in other business areas.

Overhaul and Maintenance

The projects of overhaul and maintenance include, not only overhaul of process plant or of large-scale machinery, but also maintenance of factory buildings and refitting of ships. With this type of project it is often possible to evaluate the costs of the facilities being idle during overhaul against the cost of maintenance or repair activity. In such cases, the network can normally be used a number of times, according to the need to repeat the operation at specific periods of time. With ship-refitting, for example, it may be possible to use the same network for several different ships.

Generally, one of the characteristics of this type of project is the short overall duration, and thus the short duration of the individual task to be performed. Consequently, the control system must be geared to reacting on a daily basis. Resources have to be balanced and the flexibility of overtime working is limited. However, the definition of the tasks as well as their duration and the resources required will be reasonably accurate, since there are usually historical data available.

Research and Development

The second major area of application—in research and development—includes the development of weapon systems. Also included would be the research done by companies to produce certain specified technical advances on current products, where the difficulty is to define the tasks involved and the relationships and durations of them. Specifically, a number of tasks recycle if the outcome is not successful—e.g., when a test to determine whether a certain stage has been reached proves negative. In the original development of networks, this difficulty was overcome by using three-time estimates, allowing, at worst, several cyclings of a task before completion was reached. Usually, in such cases, the control is more on times at which certain critical tasks shall be complete, than on the costs incurred during the course of the project. In fact, because these projects are difficult to control, it is advisable to define target-dates for the completion of major stages.

There will be many re-analyses of the network as results of various tasks are monitored. However, the application of networks to research-and-development problems does provide better results in timing of outcomes, when the human aspects have been successfully approached. Scientists involved in research and development may not be willing readily to submit to the external discipline of network application, but, once it is accepted, they can make a substantial contribution to the necessary technical requirements.

Construction

A third area of application is the construction industry and a whole range of projects involving construction. These include the building of factories, warehouses, and office-blocks, the setting-up of process plants, and even shipbuilding. Generally speaking, such projects are carried out by one firm on behalf of another, and the client firm wishes to know when the project or service will be ready, so that their own programmes can be completed. The firm of contractors, wishing to control the work to a minimum cost (given the constraint of a date for completion), may set themselves certain completion times for tasks within the project; but as far as the client company is concerned, what matters is the availability of the finished product. In these projects, the control of rate of expenditure provides a reasonable measure of progress, and the technique labelled PERT/COST provides a means by which planned expenditure can be evaluated to compare with actual expenditure. With networks, the cash-flow of projects can be estimated accurately and thus investment appraisals are based on more realistic information (see *Capital Investment Appraisal*, by I. W. Harrison, a companion volume in this series).

Other areas in which network analysis can provide management with better facilities for planning and control are those projects which involve either the installation of new equipment or facilities, or the launching of new products or services. (The complexities of planning the products to be manufactured are explained in *Essentials of Product Planning*, by P. Gorle and J. Long, also in this series.) As with construction, it is probably just as important to achieve deadline dates for completion of one or two critical tasks, as to be involved in particularly close control of the costs. Dates are important, because they tend to mesh in with the external requirements of the company concerned. Today, networking is used for such diverse projects as major surgical operations, planning the regular issue of catalogues, and preparing annual accounts. In many developing countries, networks are used by international agencies for the introduction of new methods, and the related training of the local population.

New Equipment, New Products, and New Services

Management of business and government enterprises are faced with many alternative uses of available finance. These (and their pay-off) initially can be evaluated by the normal procedures within the organization. When a short-list has been determined, it is worthwhile insisting that each project should be networked, at least in outline, to obtain a more accurate evaluation of the actual timing of events and costs that will be incurred. In all instances, such a diagram in itself may provide more accurate insight, as well as a basis on which to evaluate any subsequent changes in the timing or cost of operations. If the broad diagram is shown to be in error during the detailed planning stage an immediate report can be made to the senior managers concerned, who are then able to make decisions on the basis of the new results expected on comparison with previous results and with other projects.

SETTING THE OBJECTIVES

When a particular project has been formally accepted, the first step in drawing a network is to state the objectives explicitly. Networks are representations of plans to achieve objectives: the objectives must be set before carrying out the analysis by network.

For example, if it has been decided that a new oil-refinery is required, the objectives will be to build a refinery of a given capacity and type, to be available on-line by a certain date. Other specific dates may be set at this stage, during formalization of the project, in order to tie in with other activities being carried out. For example, the oil-refinery storage tanks may have to be ready about six to eight months prior to the cracking facilities, so that stocks can be absorbed from the various sources.

Another example is the setting of objectives in marketing a new product. It may be that all necessary arrangements must be completed so that the product can be on the market on June 4th. But the directors wish to have the results of all test production and distribution (with final cost evaluation) before April 4th, in order to decide whether to continue the project. Resources of production would be limited to one line until after April 4th, when, at most, two lines could be allocated.

It can be seen that objectives form a hierarchy, some of which are concerned with limits of acceptability as opposed to targets of achievement. It is most important that the objectives and sub-objectives are stated explicitly, as the first step in preparing for network analysis. Some (or even all) of these may change because of lack of feasibility or incompatibility, but they do provide the first target for manipulating the network.

Checklist

Do you have projects of overhaul and maintenance, research and development, construction, installation of new processes or products? If so, you should be using network analysis for planning and control.

The first step is a clear and explicit statement of the objectives of the project. Can you be explicit without being specific? What is the actual outcome expected? When, where, and at what cost?

Are there any restrictions on timings or resources? Once these are specified, then preparations for a network diagram can be made.

3
Preparing for the network

A network is a methodology which provides a model of the tasks necessary to fulfil the project objectives. Therefore, unless the objectives are defined at the earliest possible stage of planning, the model, represented on paper as a network, will not accurately reflect the project itself.

Once the objectives have been defined, the first network should be drawn. Even though there is little or no accurate technical information on which to base an assessment of the method of completion, some systematic estimation of technical aspects, time, and costs should be made. As the project develops, so the network can be expanded to provide a more detailed and accurate representation of the method by which the objectives are to be achieved.

Even if a project is already under way without a network, it is still possible and advisable to draw one—in this case, starting from the current situation and representing the means of completion.

MANAGEMENT OF THE PROJECT

There will be a time, near the beginning of a project, when a formal decision is taken to allocate staff and provide resources to progress the work. At this stage, the objectives will have been formulated and the first broad network may have been sketched and analysed by senior management. The management structure for progressing the project should now be specified.

Types of Structure

Some companies, particularly contractors and those engaged in industries such as shipbuilding, are already organized on the basis of project management. Here, the responsibility is assigned to a project manager, who heads a team of staff comprising the skills necessary. In such a team, a specialist in network analysis should be included to carry out the detailed drawing, analysis, and control of the network, while all other members of the team must also be familiar with the basic principles of the technique. In this type of company, according to the size and complexity of the tasks in hand, a project manager and a team may be engaged upon several projects. In these circumstances, it is advantageous to plan the work of each person so that he reports only to one of the project managers over a reasonable period of time. However, if the job is big enough, project-oriented companies and those which are organized functionally can appoint a full-time team under a manager with overall responsibility.

Organizational difficulties arise with moderate-sized projects in a functionally organized company. A full-time manager may be appointed, even though the team may be required only at intervals—reverting to their functional responsibilities at other times. Thus, there may be friction, particularly if departmental interests are in conflict with those of the projects.

The Team Approach

A compromise might be made by assigning overall responsibility to a senior manager, or even to a director, assisted by staff to carry out the planning function. One planner (or more) would be involved in the detail of the project planning and control, with responsibility for the functional co-ordination of the team. The manager with overall responsibility would be involved on a part-time basis, but he would have authority over all managers whose departments were involved.

The team itself may be dynamic, with initially only a few key functions being represented. More staff successively become involved; some for only a short period of time, others over the full duration of the project. The tasks to be carried out define the staff requirements, and the overall manager organizes the team accordingly.

The team approach has a number of advantages: people from the departments which are to implement the project are involved in the planning, and have to discipline their thinking to specify the logic of the work and the time and resources required. However, the development of initial networks with a team may take longer, unless the task is broken down into areas of work which each member can develop independently.

In a functionally organized company, small projects can be made the responsibility of the functional manager most involved. He can call on the assistance of other departments through the normal, informal structure of the company. The use of network will provide a discipline which previously may not have been formalized, although the manager may have done some rudimentary planning.

BASIC ELEMENTS OF A NETWORK

Two basic elements are required in the preparation of a network: activities, and events. But the most important characteristic of a network is the definition of logical relationships.

Activities

The series of actions necessary to complete a project can be specified as a number of separate activities. An activity embraces all the actions necessary to perform a specified task. For example, building a prototype weapon is an activity; assembling a nut and bolt is also an activity. In order to carry out such activities, resources of men, money, materials, and/or machines have to be applied. Some activities, however, may consume time only, and may not require resources—the activities like: 'Wait for cement to harden' and 'Obtain bought-out parts'.

Events

An event occurs at a unique point in time. It is usually associated with an activity to delineate a definable achievement within the completion of the project. The beginning and end of an activity are events; so also may be the moment of time when an activity reaches a given proportion of its completion—e.g., when one has built half a 100-foot-long wall.

By means of these two basic elements, a project can be planned to achieve the objectives. Objectives to complete certain stages of a project by particular dates, can be specified by reference to a particular event. Plans can be drawn up on the basis of the activities necessary to achieve the objectives, and the resources required for individual activities determined.

Logical Relationships

But the major advantage of network analysis is not in the specification of activities and events. It lies rather in the definition of the logical interrelationships between activities, and in the representation of these on network diagrams. Such interrelationships allow identification of activities which are critical to the achievement of the objectives; they permit a ready analysis of the effect of delays in some activities and of constraints in resources allocated.

The specification of the logical relationship between activities is the first stage in network analysis, once activities and events have been defined. There are two types of logical relationship: strict logic, and loose logic. Strict logic is a constraint imposed by the natural sequence of events which is impossible to alter—e.g., breakfast must be prepared before it can be eaten. Loose logic, on the other hand, represents the normal practice within a project, usually self-imposed. Thus, it may state as a logical relationship that, for example, the family get dressed before they have breakfast.

The specification of logical relationships is independent of time and resources available. At this stage, all constraints except logic are ignored. It is only after the preparation of a network diagram that time and resources are considered so that the network can be analysed. Then, if changes in the plan are necessary to achieve the objectives, decisions can be made to alter the logical relationships, the activity durations, and/or the resources allocated.

HIERARCHY OF NETWORKS

As already stated, the first network should be drawn soon after the project has been defined and the objectives set. This first network might break down the overall project into as few as ten major activities, with their logical interrelationship specified. For example, a project to build a fertilizer factory was first broken down into the main areas: milling plant; phosphoric-acid plant; fertilizer plant; packaging plant and warehouse; handling equipment and bulk storage. The overall duration and cost of these broad activities was estimated, and the overall time and cost of the project was able to be determined.

The first network might be prepared by those who are developing the project idea, and it could be presented on one piece of paper. However, once a project-team has been formally appointed, even though it is the nucleus only of the total team involved eventually, the members' first task is to review any initial network and even to question the objectives.

An analysis of the information required should be carried out, as it may be necessary to obtain it in new forms. This analysis is critical to the success of a project controlled by network, and depends on the expertise of the team. Initially, there may well be wasted effort in obtaining subsequently unwanted items, but as experience is gained, so better information specifications will be prepared.

When the first network has been finalized, each major activity can be subdivided into a detailed individual network, taking account of any interrelationships between the major tasks. For any project of moderate to large size, these second-level networks would also be broad in character, requiring the representation of between, say, 100 and 400 activities. Depending on the size of the project, there will be three, four, or even more levels of networks—breaking the activities down into further and further detail. The whole of a small project can probably be drawn as one detailed network, based on an expansion of the initial network of 10 to 50 activities. Larger projects, with further levels of networks, can usefully be subdivided into almost independent sections (defined by major tasks, for instance). The interrelationships of activities between the major tasks can be identified and relevant times fixed as sub-objectives. Figure 3.1 gives the schematic sequence of planning a project by network analysis. The arrows indicate the

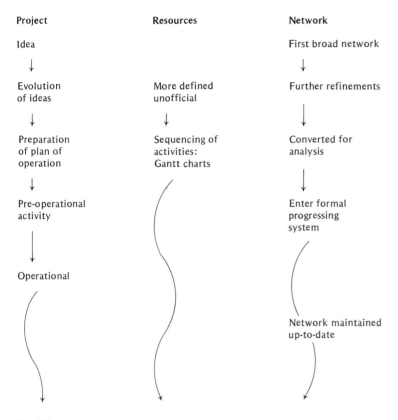

Fig. 3.1

development and use, over time, of a network for planning and controlling a project—i.e., a time-scale runs from the top to bottom.

The degree of detail into which the activities are subdivided is a matter of judgement. It is worth-while bearing in mind that the primary objective of network analysis is to plan and provide control at different levels of management. Within this objective, networks should be kept as simple as possible. In judging the value of activity subdivision, the following factors can be used:

(a) The time for control information to be prepared and presented. For example, breakdown of activities into daily tasks is not generally justified, both because of the time-lag in feeding back the information and because of the loss of flexibility at direct supervisory level. Of course, certain projects (like maintenance of plant) have been broken down into activities taking less than an hour; but these are special cases.

(b) The extent to which resources required for an activity (e.g.,

type of labour, materials, and machinery) can dictate subdivisions.
(c) The points of transfer of activities from one department or location to another.
(d) The time-consuming effort required in carrying out the subdivision and subsequent analysis. This may more than outweigh the possible extra accuracy obtained.

It is worth reiterating that the subdivision of a project into activities is cyclic. The broad network is prepared and the critical activities are identified. These are then further analysed, while the others are considered in outline only. There is no value in expending time and resources on activities which have no relevance to the overall achievement of the objectives.

ESTIMATION OF DURATION

Once the activities, events, and logical relationships have been specified, the relationship of the events to time has to be determined. Thus, it will be possible to indicate the resources required at different times, the completion dates, and the information against which actual progress can be controlled. In order to do this, the time required for each activity needs to be estimated. These estimates need only be based on judgement in the early stages of the network analysis, but as the detail is developed (and, in particular, as critical activities are identified) estimates of duration should be more accurate. In all estimation of time, bias must be avoided. It is more important that the estimates are unbiased than that they are extremely accurate.

The time estimated depends, to a large extent, on the assumed allocation of resources. For a company repeatedly carrying out similar projects, past history can be used to determine activity durations, assuming normal resources are allocated. If more accurate time estimates are required, application of the techniques of work study would provide a standard time and the resource allocations required.

In estimation, a decision has to be made on the time-unit to be used. Once this is done, then all estimates must be expressed in this unit. For projects which will take more than, say, a year, a weekly, or even monthly, unit is adequate, However, shorter projects could have activities estimated in days or even in hours. The time-unit is, of course, dependent only on activity durations; all activities should involve at least one unit, since rounding errors might become substantial if many activities are less than one.

After this stage in the development of network analysis of a project, it is possible to list the following:

(a) The activities (at whatever level of detail reached) defined as necessary to achieve the objectives.
(b) The logical relationships between the activities.
(c) The estimated durations of the activities.
(d) The resources that have been assumed necessary to complete activities in the estimated time.

A study of the work involved in building a seaside bungalow provides an example of the form of information available after the above-mentioned stage has been reached. (The bungalow-project is used throughout this book to show the practical application of network analysis.)

Mr James had just bought some land at the seaside, and he wished to occupy the bungalow during his holidays. He realized that the building might take up to two years to complete, but he wanted to take possession not later than June 1st of the following year. If the builders started on November 1st, could they finish in time?

This is typical of the problems to which the technique of network analysis should be applied. There are additionally two major reasons for using a network: the cold weather during the winter months might affect the concreting work, particularly the foundations, and the delivery periods of some suppliers might be excessive and require a tight control.

Thirty-eight separate activities were specified. The logical relationship was analysed, and the durations (in days) with associated resources were estimated. Table 3.1 lists all these factors, as developed by the builder.

From the specification of an activity listing, it is possible to prepare the network diagram, analyse the time at which each activity can and/or should start, the overall duration of the project, and a schedule of the total of each resource required over time. In practice, the list of activities may not be prepared explicitly prior to the network diagramming. Similarly, the analysis of time does not require the resource specification. But at some stage these factors will have to be stated explicitly.

Table 3.1 *Mr James's Bungalow: Activity Listing*

Activity No.	Description	Previous activities	Duration	Resources required	
				skilled	labourers
1.	Obtain joinery	—	60	0	0
2.	Obtain glass	—	20	0	0
3.	Obtain sanitary ware	—	20	0	0
4.	Obtain heating equipment	—	40	0	0
5.	Obtain electrical equipment	—	30	0	0
6.	Obtain plumbing equipment	—	70	0	0
7.	Obtain tiles	—	140	0	0
8.	Obtain bricks	—	60	0	0
9.	Obtain top-soil	—	40	0	0
10.	Survey the site	—	20	2	5
11.	Build temporary access	10	30	2	5
12.	Earthworks	11	10	2	5
13.	Lay foundations	12	30	2	5
14.	Install external drainage	6; 13	20	2	3
15.	Build shell	13	80	2	9
16.	Build boiler-house	4; 15	20	2	6
17.	Place electrical wiring	5; 15	40	2	1
18.	Install interior plumbing	14; 15	50	2	3
19.	Fix roof frames	15	20	5	10
20.	Build interior walls	8; 19	50	2	5
21.	Place roofing	7; 19	20	3	6
22.	Build outside storage and garage	15	20	4	3
23.	Install wood frames	1; 18; 20; 21	40	2	5
24.	Install sanitary ware	3; 18; 21	30	1	2
25.	Install boiler	16; 21	10	3	7
26.	Inspect wiring	17	10	0	0
27.	Lay permanent access	21	30	1	2
28.	Fit glass	2; 23	20	2	8
29.	Install radiators	18; 25; 32	30	2	5
30.	Install switches and plugs	21; 26	10	2	1

Table 3.1—*continued*

Activity No.	Description	Previous activities	Duration	Resources required	
				skilled	labourers
31.	Finish exterior	22; 27	50	1	3
32.	Surface walls and partitions	28	30	2	4
33.	Connect plumbing and test	24; 29; 30	10	2	5
34.	Clean up outside	31	10	1	4
35.	Finish interior	32; 33	30	2	5
36.	Place flagstones	9; 32; 33	30	2	5
37.	Place top-soil, and landscape	36	20	2	5
38.	Overall inspection	34; 35; 37	10	2	5
39.	Project complete	38	0	0	0

Checklist

Is your organization or company project-oriented or functionally organized. In either case, have you considered how its structure affects the management of projects. In preparing a network, what are the special virtues of the team approach?

The major advantage of network analysis is in the definition of logical relationships. Why is this more important than the specification of such hard facts as activities and events?

When a first network has been drawn, why must it be modified and expanded?

The value of activity subdivision is a matter of judgement, bearing several factors in mind. Have you determined how these factors affect your planning?

Can you explain in one, short sentence, precisely why it is essential to estimate the time for each activity?

4
Network diagram

Once the relevant activities involved in a project have been specified together with their technically logical sequence, the overall duration of the project has to be analysed.

The fundamental basis of this analysis is the preparation of a network diagram, without which not only the overall duration, but also the flexibility of the timing of each activity would be very difficult to assess. A diagram simplifies this analysis out of all proportion to the effort required to draw the network. Even if a computer is to be used, it is generally necessary to prepare a network first.

Apart from simplifying the analysis of time required, the preparation of such a diagram has additional advantages because it provides the following:

(a) Assists in checking the logical sequence of operations and any omission of activities.
(b) Ensures a discipline for the personnel involved in planning a project.
(c) Supplies a means of communication between them, and of assigning their shares of responsibility and authority.
(d) Ensures that the sequence of activities can be more readily controlled and/or modified during the course of a project.

It must be stressed again that the division of a project into activities, with the associated network diagram, is cyclic. But the techniques of preparing and analysing a network are the same, irrespective of the level of detail included.

A network diagram represents events, activities, and their sequence by a series of interlinked arrows. The diagram in which the arrow represents an activity is used in most networks today. The basic technique of drawing such a network is described in this chapter. However, it is possible to represent each activity within a box and to develop the logical sequence between the activities by arrows representing constraints. This technique has a number of advantages, and it is described in the last section of this chapter.

The basic diagram in network analysis is the arrow diagram, in which each arrow represents one activity. An arrow is drawn starting and ending with events, which are usually represented by circles. Figure 4.1 shows a typical element of a network. The circle symbol is called a 'node' within the network.

ARROWS AND NODES

Event   Event

Fig. 4.1

The arrow is pointed towards the finishing event. Apart from this convention, there is no other significance in the circles and arrows. The length of arrow has no relation to the time taken for that activity; nor is the general direction of the arrows significant, although it is usually simpler to follow the diagram if they point from left to right. When drawing a network, a brief description is usually entered against each activity. Therefore, the arrow should be sufficiently long to allow for the insertion of the description.

A network consists of a collection of arrows, linked in logical sequence according to the technical characteristics of the project activities. If, for example, activity B must follow activity A, this would be drawn as in Fig. 4.2.

THE ARROW DIAGRAM

Fig. 4.2

If there are several activities which must be completed before activity B can begin, then the corresponding arrows are drawn to end at the start-event of activity B. Similarly, all activities which cannot start until activity B is finished are represented by arrows starting from the end-event of the activity. Figure 4.3 is a typical example.

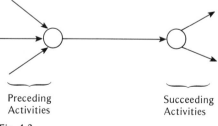

Preceding Activities   Succeeding Activities

Fig. 4.3

21

A network, consisting of arrows and nodes, represents, then, the physical sequence of activities. At this stage of the analysis, the sequence is drawn irrespective of time or resource considerations. And the diagram itself does not, in general, indicate by its scale the timing of events. An example of a small network is given in Fig. 4.4.

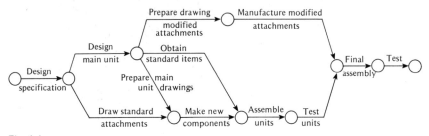

Fig. 4.4

As a check on the drawing of a network diagram, it is possible to identify activities that can be performed concurrently. These can be referred to on the actual diagram, and one may check that they are shown as concurrent. In the example given in Fig. 4.4, the activities of obtaining standard parts and making new components are such that they can be carried out concurrently. Reference to the diagram shows that these two activities have been correctly represented.

DUMMY ACTIVITIES

The method of drawing networks outlined above does not cover all the requirements for representing logical sequences.

For example, consider a simple maintenance job required on a hydraulic unit, during the overhaul of an automatic press. The activities involved are:

A. Obtain new piston-head;
B. Dismantle hydraulic unit;
C. Rebore cylinder;
D. Fit new piston-head;
E. Grind piston-head to size;
F. Reassemble unit and test.

These activities might be represented by the diagram shown in Fig. 4.5.

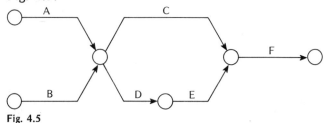

Fig. 4.5

However, in carrying out this work, it is technically possible for the reboring of the cylinder to begin as soon as the unit is dismantled, and this activity does not have to wait for the new piston-head to be obtained. But the network shown in Fig. 4.5 restricts the start of activity C (rebore cylinder) until both activity A (obtain new piston-head) and activity B (dismantle hydraulic unit) have been completed.

The difficulty is overcome by drawing a 'dummy' activity. This is represented by a broken-tailed arrow, with the direction of the arrow-head based on the same principle as that for normal activities—i.e., the tail of the arrow starts from the event which must be completed before the dummy activity can take place, and the head of the dummy activity shows the event which follows this activity. The dummy activity itself has no duration and requires no resources.

The network diagram of the activities necessary for the maintenance of the hydraulic unit should therefore be drawn as shown

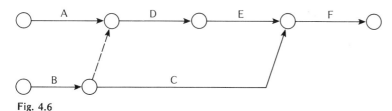

Fig. 4.6

in Fig. 4.6. Activity D (fit new piston-head) cannot begin until the dummy activity is complete, which is equivalent to activity B (dismantle unit) being complete. Yet, technically and according to the diagram, activity C (rebore cyclinder) can begin as soon as activity B is finished, irrespective of activity A (obtain new piston-head) being complete. This represents the actual logical connection between the activities.

In the project detailed in Table 3.1, the following activities are specified and numbered as follows:
3. Obtain sanitary ware;
18. Install interior plumbing;
24. Install sanitary ware;
20. Build interior walls;
23. Install wood frames.

The technical considerations are such that activities 23 and 24 cannot start until activity 18 is complete; but each one of them can start independently, as soon as activity 18 and the other associated activities (3 or 20) are complete. The diagram in Fig. 4.7 represents this sequence.

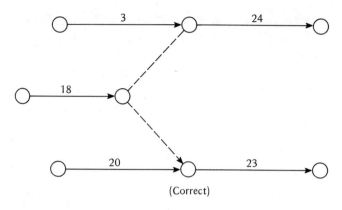

Fig. 4.7

Any diagram of these activities resembling the pattern in Fig. 4.8 would be wrong.

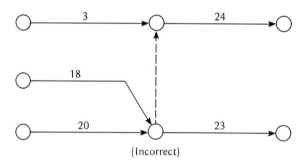

Fig. 4.8

PARALLEL ACTIVITIES AND TIME CONSTRAINTS

Many projects include two activities, the second of which cannot be started until after the first has begun, but can be started before the first project is finished. For example, the laying of foundations has to follow the trenching of the ground; but once some trenching has been done, the foundations can be started, and continued as further trenching is dug. The most rigorous way of diagramming this—unfortunately losing a little realism in the

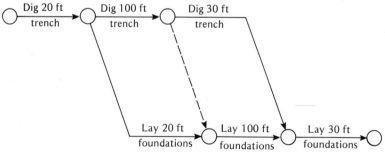

Fig. 4.9

process—is to break down the two activities into components and to plot them in sequence. Such an exercise is shown in Fig. 4.9.

Plotting on a diagram becomes complex when a number of succeeding activities may take place shortly after the start of preceding activities. Building a wall might here be included in the above example, after trenching and laying foundations, and all the activities, for reasons of further realism, may be broken down into more separate activities.

A method of plotting parallel activities is to use a 'ladder', which simplifies the drawing but reduces the rigour of the network diagram. The ladder can then be drawn with chain-linked activities, which represent time-delays only, as shown in Fig. 4.10.

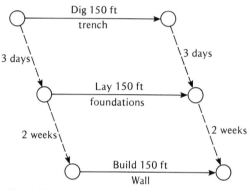

Fig. 4.10

An alternative method of showing a ladder of activities is to use a box at the end-event, instead of a circle. This is exemplified in Fig. 4.11. (The method is used by International Computers Ltd. in defining the inputs for their standard network-analysis programme.) In these ladders, the arrows representing time-delays only (or lead-and-lag arrows) need not be of the same duration. It may be technically possible to start a job within a project, say, three

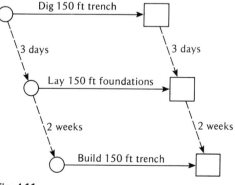

Fig. 4.11

25

days after the preceding job has started, but only to finish it ten days after the first has finished. Dummy arrows representing time-delays including no resources are often called real-time dummies. Time-delayed activities which involve no resources have, of course, to be included within the network as ordinary activities. In particular, activities involved in waiting to receive delivery of orders must be included. Other time constraints caused by factors external to the project, such as the end of the tax-year or the beginning of a new season, can also be included. These can be drawn as arrows starting from the initial event, and restricting the start of any specified activities.

EVENT NUMBERING

Once the network has been drawn, each node (event) should be numbered. Numbering has several advantages, especially if the network is subsequently to be analysed by computer. In numbering events, it is advisable to observe certain principles (although, with current computer programmes, random numbers are accepted):

(a) Event numbers at the beginning and end of an activity should be unique.
(b) The number of the end-event should be greater than the number of the beginning-event.
(c) Gaps should be left in the sequence of event numbering, to allow for insertion of activities subsequently.

In order to ensure that the number of beginning- and end-events of each activity is unique, it is sometimes necessary to use dummy activities. Consider, for example, the following sequence of activities:

A. Develop catalogue layout.
B. Obtain plates.
C. Set up type.
D. Print catalogue.

A network diagram of these activities could be drawn as shown in Fig. 4.12. However, the activity described by the pair of events 2-3 would be ambiguous. By inserting a dummy, unique numbering could be maintained; thus, the same activities as given in Fig. 4.12 could be drawn as shown in Fig. 4.13.

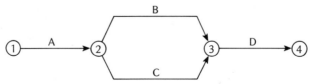

Fig. 4.12

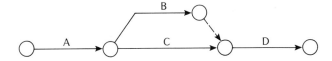

Fig. 4.13

A standard procedure, developed by D. R. Fulkerson, numbers events so that any activity has a lower-numbered event at the beginning rather than at the end. This procedure is:

(a) The initial event, with no prior activities, is numbered 1.
(b) Regard all arrows leaving numbered events as non-existent.
(c) This will leave one or more 'initial' events (with no prior activities). Label these in numerical sequence; the order in which they are numbered is irrelevant, and the numerical sequence can have numbers omitted.
(d) Repeat steps (b) and (c) until the final event, with no subsequent activities, is reached.

For a simple example of this procedure, consider the network diagram shown in Fig. 4.14. The initial event is numbered 1.

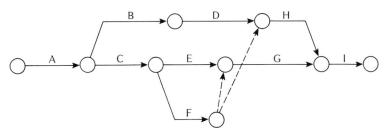

Fig. 4.14

Regarding activity A as non-existent, one initial event is left, and it would be numbered, say 10, using a sequence of numbers which are multiples of 10. The activities B and C would now be eliminated, and the two new initial events would be numbered 20 and 30. Eliminating activities D, E, and F would now produce only

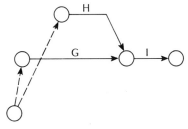

Fig. 4.15

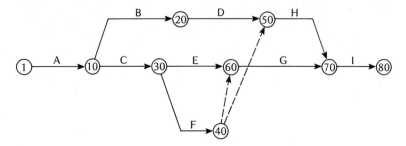

Fig. 4.16

one new initial event, since dummy activities are treated as other activities. The remaining network is shown in Fig. 4.15. This procedure is continued until all events are numbered. The resulting network is shown in Fig. 4.16.

Events are numbered to provide a comprehensive means of identifying activities. With small networks, it is possible to use letters of the alphabet to provide a reference, but with a large network this is not possible. Event numbers are useful in reporting during the execution of a project, and they are readily identified on a network diagram.

PREPARING THE FAIR COPY

The initial draft of a network is prepared by trial and error, and it is likely to look confusing. But such a draft can be redrawn neatly, observing a number of simple practices:

(a) Try to keep all arrows pointing from left to right. This can be done by inverting the positions of events or by lengthening the arrows of certain activities. (See Fig. 4.17.)

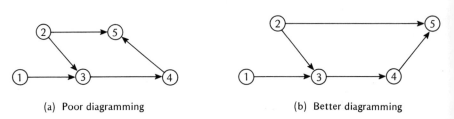

(a) Poor diagramming　　　　　　　(b) Better diagramming

Fig. 4.17

(b) Try to keep all arrows horizontal, even using 'dog-legs' if necessary. Never use curved lines. (See Fig. 4.18.) This allows better presentation of activity descriptions.

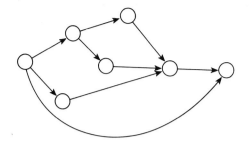
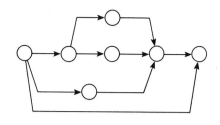

(a) Poor diagramming   **Fig. 4.18**   (b) Better diagramming

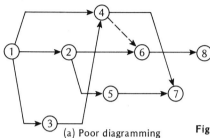
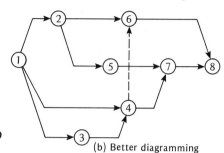

(a) Poor diagramming   **Fig. 4.19**   (b) Better diagramming

(c) Avoid arrows crossing one another, particularly the full arrows—let dummy arrows cross rather than these. (See Fig. 4.19.)

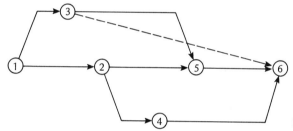

Fig. 4.20

(d) Avoid drawing dummy arrows if the sequence is already logically determined. Consider the network shown in Fig. 4.20. Event 3 must be complete before event 5, and event 5 before event 6. Therefore, the dummy activity 3–6 is unnecessary.

COMPLETE NETWORK DIAGRAM

It is now possible to draw a complete network. The network diagram of the project activities listed in Table 3.1 would be as drawn in Fig. 4.21.

Note that only dummy activities leave after activity 18, as discussed on p. 23. Note, also, that, although activity 32 is specified as being prior to activity 35, no dummy arrow is drawn, because the logical relationship is already determined by the sequence of activities 32, 29, and 33.

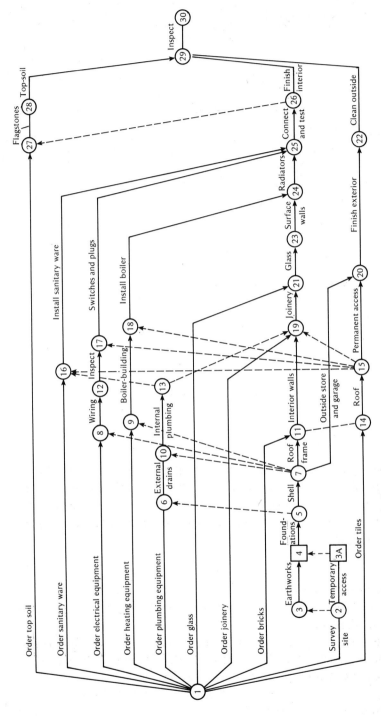

Fig. 4.21 *Mr James bungalow: network diagram*

30

A few errors inevitably arise in drawing the network, due to technical illogicalities in the original list of activities, and/or mistakes made when drawing the diagram.

POSSIBLE ERRORS

(a) It is an error to leave an event, other than the start and completion of the project, without both a prior activity and a subsequent activity or dummy activity. If such an event is logically consistent, then it should coincide with the start-project or end-project event.

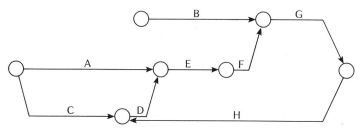

Fig. 4.22

(b) It is an error to loop. In Fig. 4.22, activities E, F, G, H, and D form a loop, indicating the absurdity that activity E cannot start until activity E is completed. When events are numbered using the Fulkerson method, the error will automatically indicate itself by your having an event which cannot be numbered.

(c) It is an error to put down an activity twice, unless it occurs twice in the project.

The method of drawing networks with arrows as activities and with circles as events was established during the original evolution of PERT and CPM. An alternative method of preparing network plans, however, was developed by Bernard Roy of the firm of

METHOD OF POTENTIALS

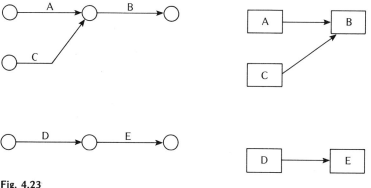

Fig. 4.23

management-consultants, Metra International, and this method has become known as the Metra Potentials Method (MPM); the method is also called precedence diagramming and activity-on-node diagramming. It is being used more and more frequently, partly because of its many advantages over the networks with arrows and circles.

In an MPM network, the activities are represented with squares or circles; while arrows represent the logical interdependence of activities. Figure 4.23 shows the basic elements of the MPM network compared with the arrow diagrams already described.

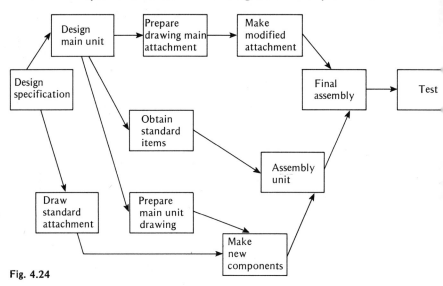

Fig. 4.24

If MPM were used to draw the diagram, the example given in Fig. 4.4 would be transposed into the diagram given in Fig. 4.24.

One of the most important advantages of a network using MPM is that it does not require dummy activities to maintain the logic. For example, the logical relationships using dummies in the network given in Fig. 4.6 (shown again in Fig. 4.25) would be transposed into the MPM network shown in Fig. 4.25. Related to the elimination of dummies, MPM diagrams simplify the drawing of parallel activities.

It is also possible to represent lead-and-lag restraints by the positioning of arrows relative to the activities. If the arrow ends on the left of the activity node, it is considered to be a starting constraint. Conversely, if it ends above or below the activity, the constraint is between the end of one activity and the end of the next. The representation of three possible relationships of activities, using both arrow diagrams and MPM diagrams, is shown in Fig. 4.26.

In most projects, efficiency is obtained by overlapping as many

Arrow diagram

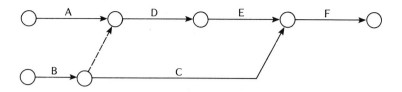

MPM diagram

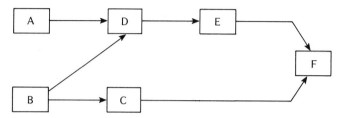

Fig. 4.25

activities as possible. Using an arrow diagram, this can only be represented by breaking up the activities into parts, while MPM diagramming does not require this subdivision. Thus, there can be considerable savings in the number of activities that have to be drawn.

The representation of activities within boxes is easier to understand than the arrows concept. It is much simpler to draw the

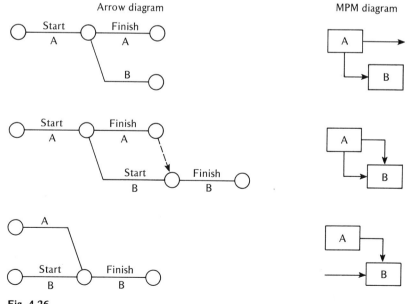

Fig. 4.26

activities on the network as they have been listed and to connect the arrows (constraints) later.

Because dummy activities are not needed on the diagram, and because activities can be entered independently of the positions of other activities, this method of preparing networks is particularly advantageous at the planning stage of a project. Any modification of the network can be done very simply and quickly without many alterations in the diagram—e.g., moving of activities, changing of events, including dummies, etc. Also, the drawing of a network, by the use of arrows and circles, tends to force a planner into consideration of ever-increasing detail. This may be a useful discipline in some cases—but more often leads to rejection of the use of networks early in a project, because of the large mass of irrelevant details.

The complete network diagram of the project to build Mr James's bungalow shown in Fig. 4.21 as an arrow diagram, would be transposed to the MPM diagram shown in Fig. 4.27. Note that the use of time constraint on activity 12 is different from the duration of activity 11. This was shown as a ladder in Fig. 4.22.

The principal disadvantage of MPM networks is that events are eliminated. Although this simplifies the network, there are many applications in which events are of importance. Also multi-project networks cannot be linked through interface events—however, it is possible to introduce a dummy activity of zero duration through which projects can be linked.

Checklist

The network diagram is drawn with arrows and circles (or boxes). Do you know what these represent?

How would you set about drawing a network, showing the logical interrelationships between activities?

Why does the arrow diagram require dummy activities and activity ladders? Does an MPM diagram require such representation? If not, why not?

When you draw the fair copy of a network, do you know how to obtain a neat and easily readable result? Do you know where to look for any errors?

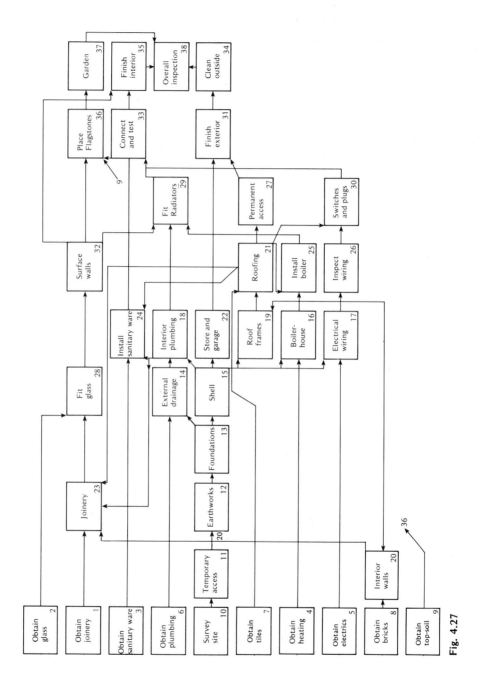

Fig. 4.27

5
Time analysis of the network

The preparation of the network diagram provides a discipline in specifying the objectives of a project and the activities involved. However, it does not enable the project manager to plan and control the activities; for at this stage he has a list only of the sequence in which they must occur. So far, there is no information on their timing, nor on those activities which must be completed on time if the project is to meet the deadline.

TIMING OF EVENTS

It is the analysis of timing that provides the manager with a schedule of events and activities. This analysis also identifies those activities which are critical—i.e., which cannot be rescheduled, in time of starting or in lengthened duration.

The non-critical activities are analysed to provide the manager with information on their flexibility of starting-times and of durations, and this indicates the degree to which their scheduling can be varied without affecting the overall time taken to complete the project.

Once this analysis of event- and activity-times has been completed, the manager can take action, before the project begins, to reduce the timings of the critical activities, thereby reducing the overall project duration. Then, during the realization of the project, he can concentrate on the critical activities, and consider only the non-critical if these become critical.

The times by which activities can start and by which activities must start, may be analysed in absolute terms by reference to dates on the calendar, or in relative terms by reference to the project start-time, set at zero. It is usually easier to carry out the calculation by the latter method and to convert to calendar-dates afterwards, if required.

Earliest Event Times

The earliest time of an event is determined by adding the duration of the activity which finishes at that event to the earliest time of the event preceding that activity, thereby determining the earliest time of the end-event (and the earliest start-time for the next activity). If a number of activities are preceding an event, then the last activity to finish, based on its earliest start-time and its duration, sets the earliest time of that event.

To illustrate the analysis of networks, Table 5.1 could stand for the list of activities of a project, together with their logical relationship and durations.

Table 5.1

Activity	Preceding activities	Duration (weeks)
A	—	2
B	A	1
C	A	1
D	A	2
E	B	3
F	C; D	4
G	D	2

The network of this project is given in Fig. 5.1. By reference to the diagram, each activity (labelled by the letter above each arrow) can be identified by the starting and finishing event-number. The analysis is then simplified by referring to these events. The duration of each activity is also entered on the diagram in brackets under each arrow.

Event 1 is assumed to occur at time 0. Therefore, event 2 can occur after two weeks (which is the sum of the earliest start of activity 1–2, added to the duration of the activity). Similarly, the earliest that events 3 and 4 can occur is three weeks and four weeks after the start of the project, respectively. Event 5 is constrained by two activities, and the earliest at which it can occur is the latest of the two times determined by these activities. Activity 2–5 can be completed after three weeks, while the

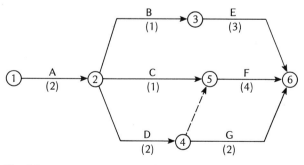

Fig. 5.1

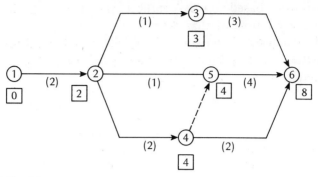

Fig. 5.2

dummy activity 4–5 can only be completed after event 4 has occurred (after four weeks). The earliest time that event 5 can occur is, therefore, after the fourth week. Event 6 is constrained by three activities, and can only occur at the end of the eighth week: which is the greatest of the completion times of activity 3–6, after six weeks, activity 4–6 after six weeks; and activity 5–6, after eight weeks. The final network, with all earliest event-times recorded is shown in Fig. 5.2.

In practice, to make the calculations easier, the results of each step of the analysis of event-times would be recorded on the network against the appropriate events. The earliest time by which the project can be completed is eight weeks, determined by the

Table 5.2 *Electronic Assembly Modification: Activity Listing*

Activity	Description	Preceding activities	Duration (days)
A	Issue design information	—	1
B	Prepare drawing of modified feed-unit	A	6
C	Design modified storage-unit	A	3
D	Prepare storage-unit drawing	C	1
E	Obtain bought-out items	C	6
F	Prepare modified scanner-unit drawing	C	2
G	Make new component	B; D	5
H	Modify feed-unit	E; G	1
J	Modify storage-unit	E; G	2
K	Modify scanner-unit	F	1
L	Final assembly	H; J; K	1
M	Test	L; N	1

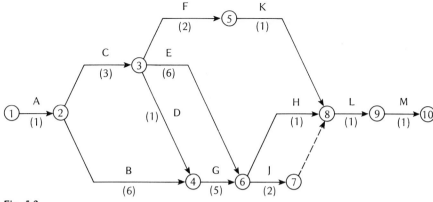

Fig. 5.3

logical relationships of the activities and the estimated duration of each.

If the events have been numbered, using the four steps described on p. 27, the event-times can be calculated in number sequence; and no errors will arise by ignoring one series of links.

The activities and durations given in Table 5.2 are those of a project to modify an electronic assembly used for process control. The network of this project is shown in Fig. 5.3.

The final network with all earliest start-times of activities recorded is shown in Fig. 5.4. The earliest time by which the project can finish is day 16, according to the logical relationships of the activities and the estimated duration of each.

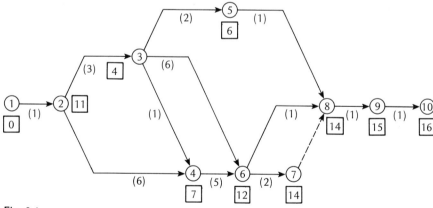

Fig. 5.4

The earliest time by which a project can be completed has now been determined. Given this time (t), it is possible to determine the latest time by which events must occur, so that the overall duration of the project is not extended beyond time t.

Latest Event Times

39

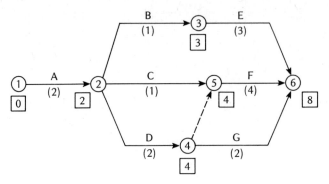

Fig. 5.5

The latest time for an event is determined by subtracting the duration of the succeeding event. If several activities are starting from the event, the least of the latest start-times of these determines the latest time at which the event must occur.

The latest start-times of each activity is calculated from the time of completion of the project. This time is assumed to be the earliest possible, and so the latest time of the end-event is the same as the earliest time. Referring to Fig. 5.5 (which is a reproduction of Fig. 5.2), the end-event, event 6, must occur at the end of week 8. The latest event-times of events 3 and 5 must therefore be week 5 and week 4 respectively (which are the latest time of the end-event less the activity durations). Event 4 must occur at the earliest of the latest start-times of activities 4–5 and 4–6. Activity 4–5 must start by week 4 (latest time of event 5 less the activity duration of zero), while activity 4–6 must start by week 6 (latest time of event 6 less the activity duration of two weeks). The latest time of event 4, then, must be week 4.

Similarly, event 2 must occur by week 2, which is the earliest of the times determined by activities 2–3, 2–4, and 2–5. Finally, event 1 must occur at time 0. The overall calculations are thus automatically checked on completion of the latest times of events,

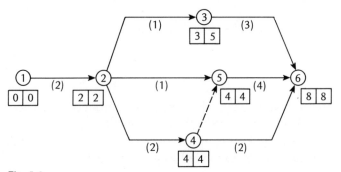

Fig. 5.6

because the latest time of the initial event must be zero. Figure 5.6 shows the final network analysis with all these results entered. Again, in practice, the results would be entered on the diagram as they are calculated.

Continuing with the project to modify the electronic assembly described in Table 5.2, the latest start-time of each activity is now calculated from the time of completion of the project. As an example of the calculations, when the latest time of event 8 has been determined, the latest times of events 7 and 6 can be determined. The latest start-time for event 7 is day 14, as the dummy arrow has no duration. The latest start-time for event 6 will be based on activity 6–7 or activity 6–8. If the duration of

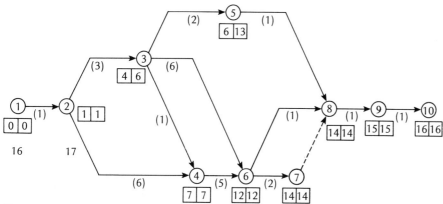

Fig. 5.7

activity 6–8 is taken from the latest time for event 8, the latest time possible for event 6 will be day 13. If the duration of activity 6–7 is subtracted from the latest time for event 7, the latest time for event 6 will be day 12. Therefore, event 6 must occur by day 12, otherwise event 7 will be delayed; and this in turn will delay the final date of completion. Event 5 need not occur until day 13 (the latest time for event 8 (day 14) minus duration of activity 5–8 (1 day)). Event 4 must occur by day 7. Event 3 must occur by day 5, which is the earliest of day 11 (from activity 3–5), day 6 (from activity 3–6), and day 5 (from activity 3–4). Finally, event 2 must occur by day 1, and event 1 by day 0. Figure 5.7 shows all these results entered.

In the analysis of the arrow and event networks, the earliest and latest event-times were determined. This does not directly provide information about the earliest and latest start- and finish-times of the activities. But with the Metra Potentials Method (MPM),

MPM ACTIVITY TIMES

referred to on p. 31, the earliest and latest start-times of activities are immediately determined.

Consider the MPM network shown in Fig. 5.8. The activities are designated by the letters A–D, and each has four boxes attached to it, used to enter earliest and latest start- and finish-times. The

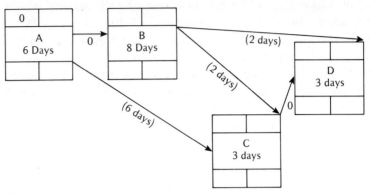

Fig. 5.8

duration entered against a constraint represents the time constraint between one activity and the subsequent activity. For example, activity C cannot start until six days after the end of activity A.

In order to calculate the earliest and latest start-times of the activities, it is useful to rank activities in much the same way as the method developed by D. R. Fulkerson. The rank of an activity is one greater than the greatest of the ranks of preceding activities. Figure 5.9 presents the rank of the activities in Fig. 5.8. The procedure for the calculation of rank will also automatically indicate whether the network has a loop in it, for the ranks cannot then be calculated.

The earliest start-time is calculated by taking the earliest time of prior-rank activities and adding the constraint time. In Fig. 5.9, activity A, the only rank-one activity, starts at a time 0. Next, taking rank-two activities (which can now be calculated), the earliest start-time for B is 6 (6 plus 0), and for activity C it is 12 (6 plus 6). At the same time, the earliest finish-time can be

Rank	Activity letter
1	A
2	B, C
3	D

Fig. 5.9

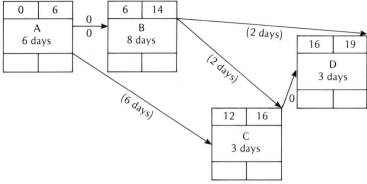

Fig. 5.10

entered. The earliest finish-time of activity B is 14 days after the start of the project. The earliest finish-time of activity C is the greater of the earliest start-time plus duration (12 plus 3 days) and the finish-time of activity B plus the constraint-time (14 plus 2). Thus, activity C cannot be complete until after 16 days. The earliest start-time of activity D is day 16, and the finish-time is the greater of this time plus the duration and the finish of activity B plus constraint. Therefore, the earliest finish-time is day 19. These results are shown in Fig. 5.10. The process is then reversed from the last activity to calculate the latest start-times. First, the ranks are reversed, and the earliest of all permissible latest start-times is calculated. The final results are shown in Fig. 5.11. From this it is possible to identify immediately the earliest and latest start-times of the activities.

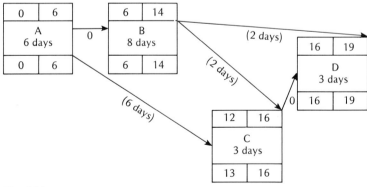

Fig. 5.11

The determination of the times at which activities can start is the first step in the analysis of a network. Do you know how to determine the earliest time of events?

Checklist

43

When the earliest time of the last event is determined, so is the overall project duration. What information does the calculation of latest event time provide?

As compared with the arrow diagram, is there a difference in the calculation of activity start-times using the MPM method?

6
Critical path and float

In the analysis of the timing of events, the earliest and latest times of some of them will be the same. Each of these events must therefore happen at a particular moment of time after the beginning of the project.

CRITICAL EVENTS AND ACTIVITIES

Such events are called 'critical', in the sense that there is no flexibility in the time at which they occur. In Fig. 6.1 (equivalent to Fig. 5.6), event 2, for example, is critical; it must happen at week 2, because, if it did not happen then, the start of activity 2-4 would be delayed beyond the latest start-time, and the overall project completion would be delayed also. Where the earliest and latest event-times are not equal, there is some flexibility between the possible earliest time and the necessary latest time of the event.

Activities between two critical events are, in general, also critical. If there is any extension in the duration of these activities, the succeeding critical event is delayed and thus the overall project completion as well. For example, activity 5-6 are critical; if the duration of the activity is greater than four weeks, event 6 will be delayed beyond the completion time.

Exceptions to this definition of critical activities are those activities which lie between two critical events, but whose duration is less than the difference between the times of the end-event and

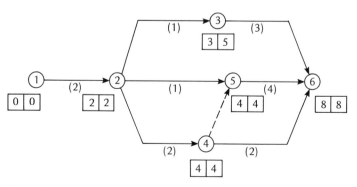

Fig. 6.1

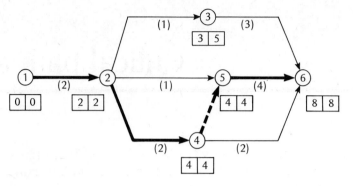

Fig. 6.2

the start-event. Referring to Fig. 6.1, activity 4–6 lie between two critical events, 4 and 6. However, the difference in time between these two events is four weeks, whereas the activity has a duration of only two weeks. This activity is not, therefore, critical.

Starting from the first event, which is always critical (earliest and latest times both zero), *a continuous path of critical activities and events* can be traced through the network. There will always be at least one critical path. There may, however, be more than one. The critical path through the network of the project shown in Fig. 6.1 would be those activities marked with heavy lines in Fig. 6.2. Note that the critical path can pass along a dummy activity.

Figure 6.3 marks the critical path of the network of a project to modify an electronic assembly. The path connects all events which have the same earliest and latest event-times. Between events 6 and 8, an analysis of the activity-times is needed to identify the critical path.

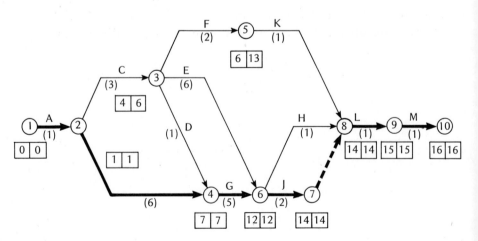

Fig. 6.3

Non-critical events have different earliest and latest times. To the extent of this difference, the finish of preceding activities or the start of succeeding activities can be delayed. Thus, each non-critical activity has flexibility at its start and/or finish.

Any non-critical activity could start at any time between the earliest start-time (EST), assuming no delay in preceding activities, and the latest start-time (LST) that is possible without delaying the overall project-completion time. The LST is determined by deducting the duration of the activity from the latest finish-time (LFT).

FLOAT

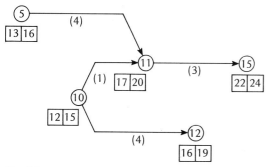

Fig. 6.4

Consider activity 10-11 in a section of the network shown in Fig. 6.4. The estimated duration of the activity is given below the arrow in the diagram. The EST is 12, while the LST is 19 (20 less 1-day duration). The activity can start at any time between 12 and 19, and, having started, the activity duration can be extended up to completion by the latest finish-time. The activity could finish as early as 13, which is the earliest start-time plus the duration: this is the earliest finish-time (EFT). In order to show the flexibility in the timing of such activities, it is useful to plot them against a time-scale. Figure 6.5 shows the possible time of activity 10-11 in Fig. 6.4.

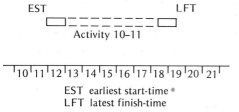

EST earliest start-time
LFT latest finish-time

Fig. 6.5

In the project to carry out the electronic-assembly modification, the network analysis resulted in the network reproduced in Fig. 6.3.

47

From this network, the earliest and latest start- and finish-times can be calculated. The results of such a calculation are given in Table 6.1.

Table 6.1 *Electronic Assembly Modification: Event Timings*

Activity	Duration	Earliest start	Earliest finish	Latest start	Latest finish
A	1	0	1	0	1
C	3	1	4	3	6
B	6	1	7	1	7
D	1	4	5	6	7
F	2	4	6	11	13
E	6	4	10	6	12
G	5	7	12	7	12
K	1	6	7	13	14
J	2	12	14	12	14
H	1	12	13	13	14
L	1	14	15	14	15
M	1	15	16	15	16

The start- and finish-times are used, in conjunction with the duration, to determine the flexibility in the timing of events and activities. This flexibility is measured by the *float* (sometimes called the 'slack') of the activities. The total float of an activity is the total range of time within which the activity can be completed without affecting the overall duration of the project. This would be any time between the earliest and latest start-time—or, of course, the earliest and latest finish-time. To state this as an equation:

$$\text{total float} = LST - EST.$$

For activity 10–11 in Fig. 6.4, the total float is seven days. The activity can be delayed in starting (and/or extended in duration) up to seven days without affecting the overall project time. But if, during the realization of a project, the total float of one activity were taken up completely, then the next activity might become critical, and the total float of some of the subsequent activities would be reduced. For example, if activity 10–11 do not start until the latest start-time, then event 11 will occur on day 20, and the total float of activity 11–15 will be reduced from four days to one day.

A measure of the extent to which the timings of activities may be altered, without affecting subsequent activities in any way, is the free float (early). This is the float available if the subsequent activities are to start at the earliest possible time. The equation is:

free float (early) = earliest time of end-event − EFT.

Similarly, a measure of the total float that is left, if a preceding event occurs at its latest time, is the free float (late), calculated by deducting the latest start-time—i.e.:

free float (late) = latest time of start-event − LST.

The free floats of activity 10–11 are shown diagrammatically in Fig. 6.6. The free float (early) is four periods, and the free float (late) is four periods. If a free float is negative, then there is no free float; the size of the negative free float provides some indication of how nearly the activity is critical.

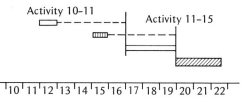

☐ Early dates schedule
▨ Late dates schedule
----- Float

Fig. 6.6

A more elaborate network and the diagram of free floats plotted against time, are given in Fig. 6.7. Using this diagram, it is possible to identify the critical path, and to read off the free floats directly. Table 6.2 summarizes these free floats.

Table 6.2 *Free Floats of Network in Figure 6.7*

	Free float	
Activity	early	late
C	1	2
D	0	1
F	1	0
G	0	1
J	1	1
K	2	0
N	2	0

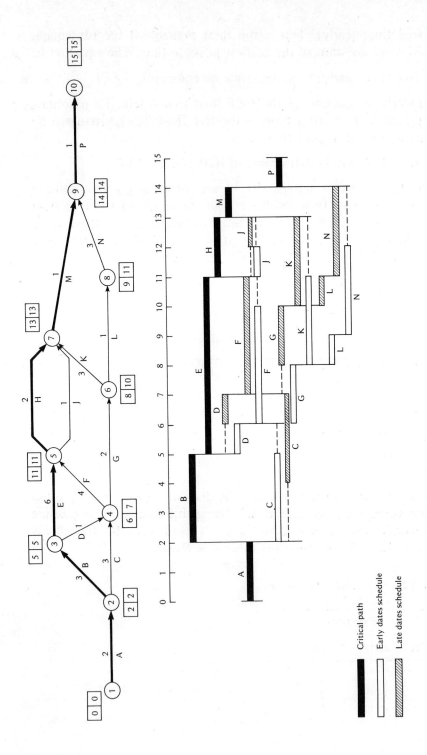

Fig. 6.7

Finally, activity float is sometimes available, which is independent of the time of the preceding event, and, if used during the execution of that activity, would not in any circumstances affect the succeeding event. This independent float is calculated by subtracting the latest time of the preceding event from the earliest time of the succeeding event, and taking away the activity duration. For activity 10–11, the independent float is one day: 17 minus 15 minus 1. This implies that activity 10–11 can be delayed or extended by one period without affecting the succeeding event, irrespective of the actual time at which the preceding event occurs within the limits of the event-times.

Table 6.3 gives the calculation of floats of the activities involved in the electronic-assembly project. These were calculated by reference to the network in Fig. 6.3 and the data in Table 6.1. The activities on the critical path have no floats and, in particular, they have no total float. Logically this must be so, and, in fact, calculation of floats is another way of identifying critical activities.

Table 6.3 *Electronic-Assembly Modification: Activity Floats*

Activity	Duration	Total float	Free float (early)	Free float (late)	Independent float
A	1	0	0	0	0
C	3	2	0	2	0
B	6	0	0	0	0
D	1	2	2	0	0
F	2	7	0	5	0
E	6	2	2	0	0
G	5	0	0	0	0
K	1	7	7	0	0
J	2	0	0	0	0
H	1	1	0	1	0
L	1	0	0	0	0
M	1	0	0	0	0

DETERMINING MPM CRITICAL PATH

In the analysis of the MPM network, the start- and finish-times of each activity are calculated in order to prepare time-estimates. Therefore, the critical path and the floats can be determined directly from the data available. The critical path through the MPM network follows those activities with identical earliest and latest start- or finish-times. In fact, this is equivalent to identifying

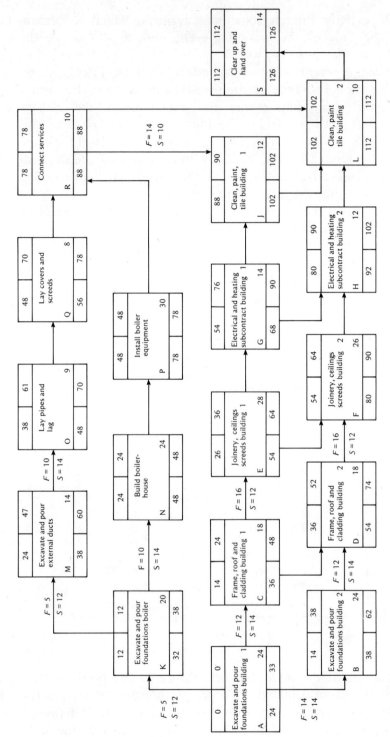

Fig. 6.8 *MPM Diagram of Building Contract.*

activities with no total float. Referring to the MPM network shown in Fig. 5.11, the critical path is along activities A, B, and D.

However, in more complex networks with many parallel activities, the identification of the critical path and the determination of floats require careful interpretation. Figure 6.8 shows an MPM diagram with a number of parallel activities, constrained on starting and finishing times. It is interesting to note that finishing constraints represent the fact that the activities cannot finish earlier, whereas, in arrow diagrams, the convention is such that parallel activities appear to be unconstrained on the time of finishing. In the network shown in Fig. 6.8, the start-to-start and finish-to-finish dependencies are shown next to the constraint arrows. In calculating finishing times, it is necessary to evaluate the finish-to-finish constrained time, as well as the activity start-to-finish time, to determine the activity finish-time. If dependency times are not given, the subsequent activity cannot start until the end of the prior activity.

The critical path through the network in Fig. 6.8 is between the start of activities A and K; between the start of activities K and N; and then along activities N, P, R, between the finish of R and J; then L and S. Thus, it can be seen that, in MPM networks, the critical path may be along a constraint (i.e., along a dummy in an arrow diagram).

The floats of non-critical activities can be determined from the activity start- and finish-times calculated. However, the floats of parallel activities must be considered differently from the floats of non-parallel activities. In fact, a series of parallel activities have to be considered as one activity, if the total float and free floats are to have the same interpretation as the floats of non-parallel activities.

The floats of parallel activities often constrain scheduling to changes only in the duration of the activities and do not allow changes in the start/finish times. For example, activity K *must* start on the twelfth day, though it can take any time between its scheduled duration of 20 days and the extended duration of twenty-six days. Similarly, activity J can start between day 88 and day 90, but it cannot finish before, and must finish by day 102 (constrained by the finish-time of activity R).

For parallel activities not associated with a critical constraint, the same problem of interpretation of floats is encountered. Consider activity C in the MPM diagram shown in Fig. 6.8. The total float would appear to have two values according to the start-times or finish-times (it could be considered as the lesser of the two values, i.e., ten days). The free float (early) cannot be determined directly for such an activity; it could be interpreted as

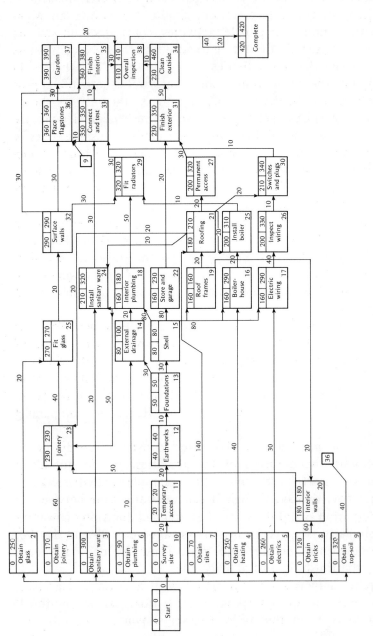

Fig. 6.9 *Mr James's Bungalow: MPM Network Analysis.*

the degree of flexibility in the duration—i.e., extending the 18-days actual duration to the constrained 22-days required.

In most practical applications of networks, only the total floats and the free floats (early) are used in planning and controlling a project. An MPM network, constrained on start-times only, can be analysed by determining the earliest and latest start-times. From these two, the critical path, and the total floats and the free floats (early) can be determined.

Table 6.4 *Mr James's Bungalow: Floats*

Activity No.	Total float	Free float (early)	Activity No.	Total float	Free float (early)
1	170	170	20	0	0
2	250	250	21	30	0
3	300	190	22	170	50
4	250	120	23	0	0
5	260	130	24	110	110
6	90	10	25	110	110
7	70	40	26	130	0
8	120	120	27	120	0
9	320	320	28	0	0
10	0	0	29	0	0
11	0	0	30	130	130
12	0	0	31	120	0
13	0	0	32	0	0
14	80	60	33	0	0
15	0	0	34	120	120
16	130	20	35	20	20
17	130	0	36	0	0
18	20	0	37	0	0
19	0	0	38	0	0

The project described in Table 3.1, and drawn as an MPM diagram in Fig. 4.27, can be analysed on the basis of start-to-start constraints, generally equal to the activity duration. The critical path and the floats can then be calculated. Figure 6.9 gives the completed network, with all constraints and earliest and latest start-times identified. The floats are then calculated and the results are tabulated in Table 6.3. The critical path (with no total float) can be seen to be the following activities:

10. Survey site;
11. Temporary access;
12. Earthworks;
13. Foundations;

15. Shell;
19. Roof frames;
20. Build interior walls;
23. Install wood frames;
28. Fit glass;
32. Surface walls and partitions;
29. Install radiators;
33. Connect plumbing and test;
36. Place flagstones;
37. Place top-soil and landscape;
38. Overall inspection.

With all the analyses now available, action can be planned so that the project is completed on time. During the project, the network should be recalculated, as delays in non-critical activities may affect the critical activities, indicating the need for a switch of concentration by the project manager.

Checklist

Are you clear about the definition of critical activities and events? What is the characteristic of the earliest and latest times of critical events?

A critical path, consisting of a continuous chain of critical activities, can be traced through a network. Is there *always* such a path?

Non-critical activities have (in their timing) a flexibility, which is shown by the possible activity starting-times. What are EST, EFT, LST, and LFT?

Using these and the earliest and latest event-times, it is possible to calculate floats. Do you know the difference between total float and free float (early)? What are the important relationships between these floats and subsequent activities?

What does a total float of zero indicate?

Is there a critical path through an MPM network?

7
Extensions of the analysis

The basic concept of network analysis has been outlined in chapters 3, 4, 5, and 6; but the technique provides additional information and facilities.

In the analysis of networks, all times were related to a start-time of zero. However, in practice, it is often necessary to prepare a schedule of events based on calendar-dates. Indeed, it may be necessary to fix some of the events on particular dates in order to co-ordinate with other external projects or actions. If so, the network is controlled by these events, and the analysis has to allow for their timings.

Another factor which has been taken into account in the development of networks is uncertainty in the estimation of activity durations. Also, the effects on cost, resulting from reducing the activity duration, can be considered by application of the critical-path method.

EVENT DATES

Once the network has been analysed relative to the starting event, it is possible to schedule the activities. This can be done by placing a relative time-scale against the calendar-dates, so that each event can be given earliest and latest dates.

Alternatively, the earliest and latest time for each event can be identified by direct reference to calendar-dates. In converting from relative to calendar times, however, care must be taken in specifying the actual working days available. For example, a standard of five working days each week can be used. The dates of any national holidays must also be specified so that they are not included in the consideration of working time.

If durations have been worked out as weekly periods, it is customary to use a relative scale based on week-numbers through the year. With such a network the durations are likely to be sufficiently long to counteract the effect of holidays; but if an activity of short duration is likely to occur over a period covering several holidays, one must ensure that the duration is extended (or dates of events altered) to allow for this.

Consider the project network analysed in Fig. 5.5. The calculated start-time of the activities is based on the number of days after the project begins. The activities could then be listed with absolute dates, as shown in Table 7.1.

Table 7.1 *Electronic-Assembly Modification: Schedule*

Preceding event	Succeeding event	Activity	Earliest start	Earliest finish	Latest start	Latest finish	Total float
1	2	A	Mar 16	Mar 16	Mar 16	Mar 16	0
2	4	B	Mar 17	Mar 24	Mar 17	Mar 24	0
2	3	C	Mar 17	Mar 19	Mar 19	Mar 23	2
3	4	D	Mar 20	Mar 20	Mar 24	Mar 24	2
3	6	E	Mar 20	Mar 31	Mar 24	Apr 2	2
3	5	F	Mar 20	Mar 23	Apr 2	Apr 3	7
4	6	G	Mar 25	Apr 2	Mar 25	Apr 2	0
6	8	H	Apr 3	Apr 3	Apr 6	Apr 6	1
6	7	J	Apr 3	Apr 6	Apr 3	Apr 6	0
5	8	K	Mar 24	Mar 24	Apr 6	Apr 6	7
8	9	L	Apr 7	Apr 7	Apr 7	Apr 7	0
9	10	M	Apr 8	Apr 8	Apr 8	Apr 8	0

In this schedule of absolute dates, note that the weekends have been omitted in the consideration of working days, and that the Easter-holidays have a direct effect in extending the absolute time of the project. Thus, although only sixteen working days have been scheduled, the total duration between start and finish is twenty-four days.

EVENT-CONTROLLED NETWORKS

A project may have to be planned so that some events coincide with certain fixed calendar-dates. The ceremony of laying a foundation-stone, for example, may have to be arranged several weeks in advance, and the event must then be given a specific date. By fixing such an event-time, which is implicitly the latest time for that event, it is possible, *either* that there is no critical path through the network up to that event, *or* that there are some paths with negative float. The negative floats indicate the activities which must be shortened in duration if the event-time is to be achieved.

The network shown in Fig. 7.1 represents the activities necessary to carry out a surgical operation. (Event 13 was left out, as it was considered unlucky to end the network of an operation with that number!) The times are given in hours and there are twenty-four hours available for schedule each day. The network starts with the preparation of the patient and of the equipment for the operation,

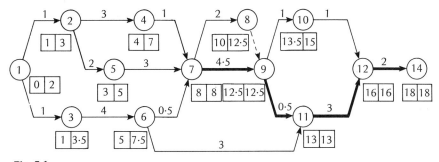

Fig. 7.1

and it ends with the post-operation tests. The time at which the operating theatre was available had been predetermined, and therefore the time for event 7 was fixed.

Initially, the operation was not thought to be an emergency, and the theatre was booked for eight hours ahead. The event-times were calculated on this basis. The surgeon, however, wished to leave as soon as possible after completing the post-operation tests. Figure 7.1 shows the event-times for the network and there is no critical path prior to event 7. The calculations are carried out in event-number order; and on reaching the event fixed in time, the earliest time is reset to the fixed time, thus saving a double-calculation of subsequent event-times.

The surgeon considered the timings and insisted that the operation be brought forward. Another operating theatre was

Table 7.2 *Surgical Operation: Schedule-2 Float*

Preceding event	Succeeding event	Duration (hours)	Total float
1	2	1	−1
1	3	1	−0.5
2	4	3	0
2	5	2	−1
3	6	4	−0.5
4	7	1	0
5	7	3	−1
6	7	0.5	−0.5
7	8	2	1.5
7	9	4.5	−1
6	11	3	2
9	10	1	1.5
9	11	0.5	0
10	12	1	1.5
11	12	3	0
12	14	2	0

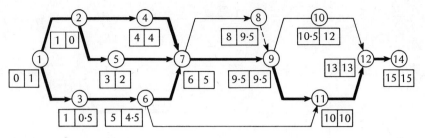

Fig. 7.2

available, but it had to be released within ten hours for a subsequent operation. Therefore, the latest time for event 11 was fixed at ten hours. The event-times were recalculated. The times are shown in Fig. 7.2; the total floats are given in Table 7.2.

Given the results in the table, the surgeon could consider action which would reduce the duration of the activities with negative float, so that the critical event, number 11, would be reached in time.

UNCERTAINTY OF ACTIVITY TIMES

The activity times used in order to determine event times and overall duration of projects are always subject to variation in practice. This may be due to variations in the measurement of time required, particularly for research work, and/or due to the external environment—e.g., building-workers may be delayed by bad weather.

The uncertainty of the estimated activity times was one of the major problems facing the US Navy when devising PERT. To allow for such uncertainty, the probability of different durations for each activity was evaluated. It was assumed that the probability of an activity's taking a particular time would be of the form shown in Fig. 7.3.

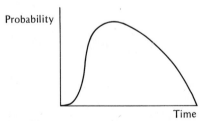

Fig. 7.3

To determine the average time for an activity and the spread of possible times about that average, three estimates of duration are used:

(a) The maximum time, if all conceivable delays occur (t_{max}).
(b) The most likely time (t).
(c) The minimum time, if no delays occur (t_{min}).

The average time of the activity is then assumed to be

$$\frac{t_{max} + 4t + t_{min}}{6}$$

which is a reasonable approximation to the average of the form shown in Fig. 7.3. The spread of possible times is described by a proportion of the range ($t_{max} - t_{min}$).

The network is then first analysed, using the calculated average time for each activity. Each event will have an associated earliest and latest time, and each of these will be subject to possible variation due to the uncertainty of the activity times. By taking account of the range of estimates for each activity, a range of possible event times can be determined, and the probability of any particular event time could take the form of the curve shown in Fig. 7.4.

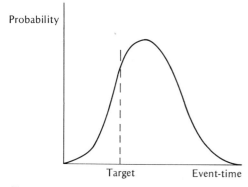

Fig. 7.4

The probabilities of achieving any target event time can then be weighted against action necessary to reduce the average duration of preceding activities and/or their range of possible durations. Such an analysis of uncertainty is particularly relevant to large projects involving research and development, and is not applied often in projects with known technical content.

MULTI-PROJECT ANALYSIS

Within an organization, several projects may be undertaken simultaneously and, if each of these is independent, then the network analyses can be carried out for each project without reference to the other projects. However, such projects may well be interdependent, constrained by external factors or resources. With network analysis, the interdependence can be taken into account and the projects planned in accordance with such constraints.

The first step is to identify the interfaces between the projects, so that common events can be included in all relevant projects when they are being drawn. When drawing out the diagrams, the interface events should be clearly identified and preferably given the same number in all the diagrams. A record of these interface events might be kept to improve control during the process of the projects, so that the effect on the projects of any changes in one project can be determined.

The common interface events are also used to plan large projects with many subcontractors. It is possible to break down such projects into a series of sub-projects (sub-nets) with clearly defined interface events. Then, each sub-project can be drawn up and analysed separately. Finally, the effect of one sub-project on the others can be determined and the final schedule of interface drawn up. A skeleton network of the interface events and the 'activities' between them can be drawn and used to plan and control the total project through the interface events and any milestone events included. The individual sub-nets will have still to be drawn up in detail, but this is much simpler—and probably more accurate—than attempting a large network without subdivision.

CRITICAL-PATH METHOD

The Critical-Path Method (CPM) extends the straightforward analysis of a network. The objective is to evaluate the extra costs of reducing the time taken by activities. These extra costs are then compared with the savings due to reduction of the overall project time. The technique was used originally to evaluate alternative schedules of maintenance and overhaul in large process plants—where production in down-time can be directly related to reduction in costs.

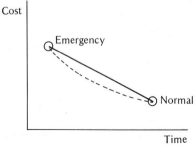

Fig. 7.5

For each activity, an estimate of the normal duration and cost is made. A second estimate is then calculated of the duration of the activity and the associated cost, if emergency action is taken. It is then assumed that there is a linear relationship between cost and duration, lying between these two extremes, as shown schematically in Fig. 7.5. In practice, the costs are more likely to take the form shown as a broken line in the figure.

The first stage in CPM is to analyse the network, using the normal durations for all activities. The completion date is thus calculated, and provides the datum from which savings can be evaluated. The next stage is to attempt to reduce the overall duration by shortening some of the activities—the first to consider being those on the critical path. However, as these are shortened, activities on other paths may become critical, and they will have to be shortened at the same time as the previous critical activities. As the duration of the project is made shorter, therefore, more activities have to be shortened simultaneously, and so the rate of increase of costs will accelerate. This rate of increase is shown schematically in Fig. 7.6.

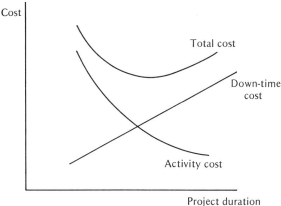

Fig. 7.6

At the same time, the cost of plant out of action will reduce. This reduction is usually linear over the range of durations considered (as shown in Fig. 7.6). Combining these two cost elements, there is a total-cost curve which has its minimum at some duration. This, then, would be the planned project duration, and the activities would be listed with the relevant degree of emergency action required. The actual execution of such a plan is more complicated than one based on single 'normal' activity times, as there are likely to be many critical paths through the network. Thus, there will be many critical activities, all to be controlled within the 'emergency' time.

CPM has been used in large-scale projects, particularly plant and building construction, where reasonable time and cost estimates are available and where the effect of changes in overall duration can be measured. If the fixed costs over the duration of the project are large, the main objective (to reduce the time taken to a minimum) is sufficient, and elaborate evaluation is unnecessary.

GANTT CHARTS Once a network has been prepared, the corresponding Gantt chart can be drawn. The interpretation of a network on a Gantt chart is done by placing a bar to represent each activity. The timing can be within the float available, but care must be taken to ensure that, if some of the total float is absorbed, subsequent activities are shifted correctly. The Gantt chart is generally based on timing activities at their earliest starts. For the project network shown in Fig. 5.5, the corresponding Gantt chart, based on earliest start-times, is shown in Fig. 7.7.

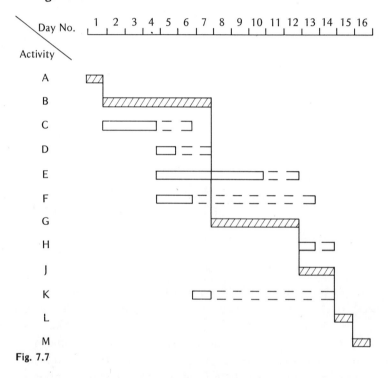

Fig. 7.7

On the same chart, total float has been represented by using some form of broken bar. The critical activities are then identified by the lack of any float. These can be marked by shading. It is useful to join the critical activities by vertical lines to allow ready recognition of the interconnections between them.

The technique of drawing Gantt charts is in general use and only an outline has been given. The charts provide managers with a visual representation of the timings of activities which can be readily interpreted. Gantt charts are thus a useful complementary technique to network analysis and should be used whenever a schedule is drawn up for presentation to managers.

TIME CONSTRAINTS

When preparing some networks, it may be helpful to use an activity which is only a time constraint on certain events. This is similar to the event-controlled network in which certain key events have specified times at which they must occur. Time-constrained activities are used when some external factor might affect the timing of events. For example, it may be considered necessary that certain activities do not occur until after a national Budget. From the start-time, this would be 87 days ahead; Fig. 7.8 shows how this might be represented on a network.

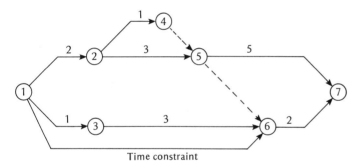

Fig. 7.8

Checklist

What are the complications of placing calendar-dates on the relative timings of events and activities?

Of course, once dates are determined, the network can be controlled on the more important of these, but if an event occurs later than required, can it be brought forward in time? If so, what would be the effect on critical and sub-critical paths?

In planning with network analyses, why are Gantt charts sometimes used as well?

8
Resource allocation

In the analysis of a project network, the overall project time and the critical path are first determined on the basis of the logical relationships between the activities involved and the estimated time to complete each activity. But there is uncertainty in the estimation of time; and, to a large extent, the resources allocated to an activity determine the time it will take. Thus, the availability of equipment, staff, and/or money can be crucial to the achievement of the project schedule.

Up to this stage in the planning of a project, resources have been implicitly assumed to be unlimited, so that each activity can be allocated the staff required. Even if this were so, the project manager would need to know when and how much of each resource was required in order to keep to the timetable. When preparing a schedule of resource requirements, it is worth-while attempting to level them out, period by period, within the limit of activity floats: this will reduce the costs of idle resources and simplify the administration of the project.

In practice, some or all of the resources may be exactly defined and limited. In these circumstances, the project manager must ensure that the totals of the resources required for activities occuring simultaneously do not exceed the resources available. This may involve, for certain activities, a rescheduling which may extend the overall time of the project. However, in the planning stage, an increase in the overall project duration, due to limitation of resources, is a natural extension of the network analysis. The resources available are simply further restraints. The final schedule provides the project manager with a plan, not only of activity timings, but also of the timing of resource requirements.

The methods by which activities are rescheduled (to allow for resource levelling or limitations) tend to be empirical—i.e., a rule is set for progressively improving the allocations until an acceptable solution is reached. This method becomes more complex, the bigger the network (number of activities and number of different resources); and, as *all* resources are seldom limited, an exploration of their different possible levels requires many separate calculations of the project schedule. All but the simplest networks involve so many interrelationships of activities and resources that schedules can best be prepared by use of a computer. This has a further advantage in that, during the execution of a project, the network analysis can be up dated economically and quickly.

It is possible to study the range of resource requirements, over a period of time, by counting the number or amount of each resource on all the activities occurring in each week. The summation can be based on each activity starting at its earliest time, or, at the other extreme, on all activities starting at their latest times.

TOTAL RESOURCES NOT LIMITED

Table 5.1 sets out the activities involved in a project to modify an electronic assembly. The estimated resources required to complete each activity in the time stated are listed in Table 8.1. In

Table 8.1 *Electronic-Assembly Modification: Resource-Requirements*

Activity No.	Description	Duration (days)	Draughtsmen (per day)	Engineers (per day)
A	Issue design-information	1	1	2
B	Prepare drawing of modified feed-unit	6	2	4
C	Design modified storage-unit	3	3	1
D	Prepare storage-unit drawing	1	1	2
E	Obtain bought-out items	6	0	0
F	Prepare modified scanner-unit	2	1	2
G	Make new component	5	1	1
H	Modify feed-unit	1		2
J	Modify storage-unit	2		3
K	Modify scanner-unit	1		3
L	Final assembly	1		2
M	Test	1	1	1

Note: Activity E does not require any resources; it is an activity involving time only.

order to analyse the resources required, Gantt charts are drawn, based on the earliest start-time and the latest finish-time of each activity. Figure 8.1 (equivalent to Fig. 7.8) shows the Gantt chart of the activities, each starting at the earliest possible time, the critical path being identified by cross-hatched activities.

The resources required can now be summed and plotted on histograms, one for each resource. On the first day, activity A only

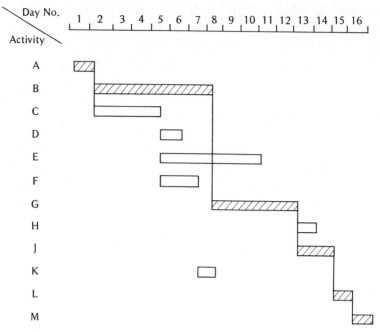

Fig. 8.1

would occur—so one draughtsman and two engineers would be required. On days 2, 3, and 4, both activities B and C would be taking place, so five draughtsmen (two for activity B, and three for activity C) and five engineers (four for activity B, and one for activity C) would be required. On day 5, four activities would take place (four draughtsmen and eight engineers would be required).

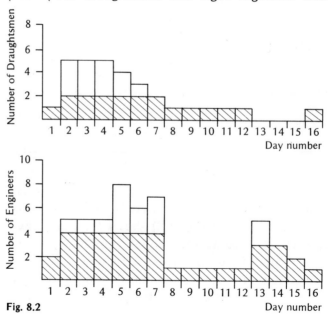

Fig. 8.2

This procedure for summing resources is continued for each day of the project plan. The final results can then be presented by means of a table and/or histogram.

Figure 8.2 shows the resulting histogram of resource-requirements, based on each activity starting at the earliest time. Those for critical activities are represented by the shaded portions. Only the resources in excess of these requirements can be rescheduled for levelling.

A method of checking that the resources have been fully scheduled, in this procedure of resource-allocation, is to compare the number of resource-periods plotted on the histogram with the calculated resource-periods from the project activity list (see Table 8.1). The calculated resource-period is obtained by multiplying the duration of each activity by the resources required for the activity per period, and summing the results. For example, the number of draughtsmen-days plotted on the histogram in Fig. 8.2 is 31, and Table 8.2 presents the calculated draughtsmen-days.

Table 8.2 *Electronic-Assembly Modification: Check-Analysis*

Activity	Duration	Draughtsmen required (per day)	Draughtsmen-days
A	1	1	1
B	6	2	12
C	3	3	9
D	1	1	1
E	6	—	—
F	2	1	2
G	5	1	5
H	1	—	—
J	2	—	—
K	1	—	—
L	1	—	—
M	1	1	1
			31

From the histograms, it is possible to prepare a schedule of the resources needed per period in order to start each activity at its earliest start-time. It is possible also to plot resources on the basis of the latest finish-time of each activity. The Gantt diagram of the activities listed in Table 8.1, based on the latest finish-time of each activity, is shown in Fig. 8.3. The histograms of the resources required for the activities based on the latest finish-times are shown in Fig. 8.4.

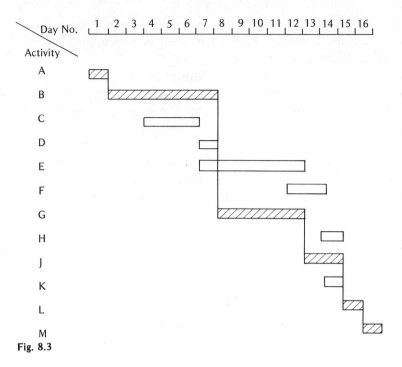

Fig. 8.3

There is a considerable difference between the timing of these resource requirements and those based on earliest start-times. Between these two extremes, there is a range of possible resource

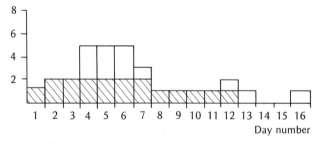

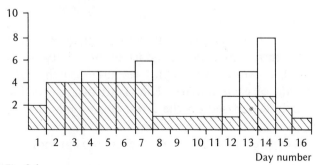

Fig. 8.4

requirements, depending on the exact time at which the activities are scheduled to start. It is by using the flexibility in the timing of non-critical activities that the resources can be levelled without affecting the overall project time. In Figs. 8.2 and 8.4, the degree of flexibility is indicated by the unshaded portions of the histograms.

RESOURCE LEVELLING

The actual resources can be optimally scheduled only by evaluation of every possible alternative. For a very small project this may be possible, but for large projects the costs, even using a computer, would be prohibitive. Yet a very good solution can be obtained by the application of reasonably straightforward rules. One approach is to use the free float, associated with the activities, in order to move resources from peak-demand periods to low-demand periods. The process is complicated, however, by the interdependence of resource requirements—e.g., in levelling one resource, another may be further peaked.

In the project to modify an electronic assembly, the resources required are shown in Figs. 8.2 and 8.4. The peak requirement for engineers occurs on day 5, based on earliest start-time, and day 14 based on latest finish-time. The activities D and F can both be varied to reduce this peak. Activity D has a free float (early) of

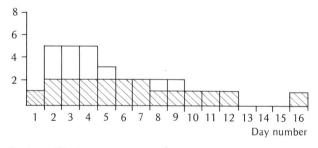

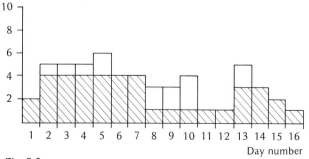

Fig. 8.5

two days, and activity F has a free float (early) of seven days (see Table 6.2). If these activities could occur between day 8 and day 12, this would lead to a substantial levelling of resources.

Moving activity D within the range of free float (early), will not reduce the peak; and, in fact, if moved to day 7 it would increase the peak requirements. So the first step could be to move activity F. If there is no criterion for choosing the periods to which the activity is rescheduled, then choose the earliest period, as any subsequent delays in starting the activity will not be so critical. Activity F would then be rescheduled to days 8 and 9; and, as a result of this, activity K must be rescheduled to preserve the logical relationship between those two activities. Thus, activity K must be scheduled to start after day 9, and can vary up to day 14. It is reasonable to schedule K to day 10. The resultant requirements of resources would then be as shown in the histogram in Fig. 8.5.

Continuing this reassessment of peaks, there will be no effect on the peak requirements if activity C or activity D is moved within its total float. Six engineers will be required on one of the days. There may be advantages in rescheduling activity C and activity D in order to provide a more gradual build-up of resource requirements at the beginning of the project.

Towards the end of a project, an increase in resource-requirements for just a short period may be difficult to administrate, and probably staff will be kept assigned so that there is no risk of

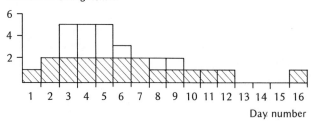

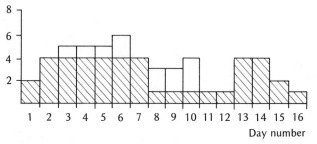

Fig. 8.6

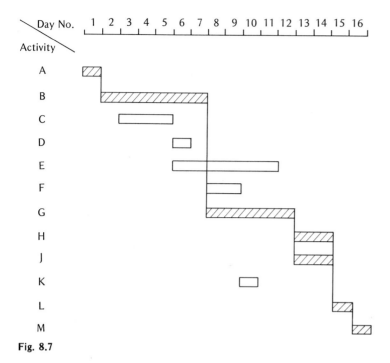

Fig. 8.7

delay. It may, therefore, be advantageous to reduce the requirements for five engineers on day 13. Activity H can be rescheduled without affecting the project completion date. But shifting the activity within its total float (one day) will not alter the requirement for five engineers. However, two engineers are allocated to activity H, and it takes one day to complete, and it may be

Table 8.3 *Electronic-Assembly Modification: Resource-Schedule*

Activity	Duration	Schedule start Day No.	Total float	Resources required	
				draughtsmen	engineers
A	1	1	0	1	2
B	6	2	0	2	4
C	3	3	1	3	1
D	1	6	1	1	2
E	6	5	1	0	0
F	2	8	4	1	2
G	5	8	0	1	1
H	2	13	0	0	1
J	2	13	0	0	3
K	1	10	4	0	3
L	1	15	0	0	2
M	1	15	0	0	1

technically possible to allocate only one engineer to the activity, extending the duration from one day to two days.

The final histograms of the rescheduled resource requirements are shown in Fig. 8.6. The Gantt chart of the activities would now be as shown in Fig. 8.7. It is interesting to note that, with the change in timing of activity H, there are two critical paths through the network: one through activity J, and the other through activity H.

Finally, all the requirements and plans will be listed for the project manager, so that he can control the actual project to achieve the desired objectives. Table 8.3 shows a typical list, based on relative times.

RESOURCES LIMITED

In the levelling of resources, no limitation was imposed on the total resources that were available, and the only criterion was reduction of peak requirements. If the resources were limited, it would be necessary to continue rescheduling the activities, beyond their total float, if necessary, to reduce the resource requirements below the limits.

In the final schedule of resources, shown in Fig. 8.6, five draughtsmen and six engineers were required for at least some part of the project. Now let us suppose that four draughtsmen only are

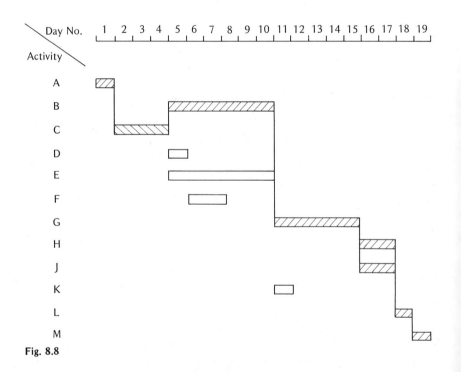

Fig. 8.8

Number of Draughtsmen

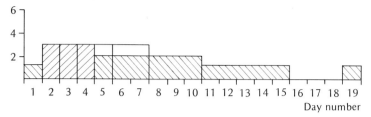

Number of Engineers

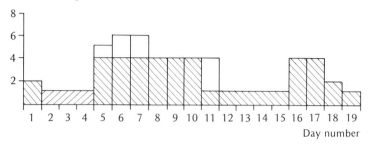

Fig. 8.9

available. Then activities B and C must be rescheduled so that they do not occur simultaneously. If activity B were rescheduled to start on day 5, shifting back activity C to the earliest start-time, the total project would be extended by three days. The resulting Gantt chart of activities and the corresponding histogram of resource requirements are shown in Figs. 8.8 and 8.9, respectively. Conversely, if C were rescheduled, the project would be extended by four days. Therefore, this alternative was not considered further.

Activities D, F and K have also been rescheduled to provide some smoothing of the resource requirements. In fact, activity F could be started on day 11, and so reduce the peak requirements for engineers to a maximum of five.

In the example, the draughtsmen were limited to four, but they were considered to be available at any time planned. It is also possible to allow for varying resources available over a period of time, and to reschedule the activities in relation to limits imposed at each stage of the project.

Apart from rescheduling the activities, in their entirety, to a different starting time, it may be possible to extend or reduce the time taken by means of allocating less or more of the resources. This can become complicated when, for example, one unit of resource is required for two weeks and it would be useful to extend the time to, say, four weeks. The rescheduling of an activity by splitting may be possible. For example, part of an

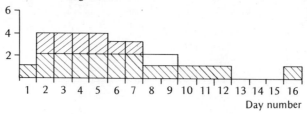

Fig. 8.10

activity having taken place, the resources would be rescheduled to a more critical activity so, once completed, the rest of the original activity could be finished. Similarly, it may be possible to split resources. For instance, in the electronic-assembly project, activity C might be extended in duration by allocating only two draughtsmen for four weeks and one draughtsman for one week. The problem then arises: What of the single engineer? Would he be required for the full five weeks, or could he be scheduled to be available for only the three weeks originally estimated?

Figure 8.10 shows the histograms of resource requirements based on the splitting of resources and the rescheduling of activities as appropriate. This extension of activity C has again created two critical paths through the network, one through activity B, and the other through activities C and D.

MPM RESOURCE ALLOCATION

The approach outlined in this chapter is also used to allocate resources in an MPM network. This approach is adequate for a manual appraisal of small networks with only a few resources, but with larger networks and/or numerous resources, the amount of calculation involved is extended considerably. For example, it is necessary, each time an activity is rescheduled beyond its free float, to recalculate the earliest start-time of subsequent activities and the resources required. Also, if the objective is to determine the most suitable levels of the different resources, many separate calculations of the network would have to be carried out. Such problems can best be solved by using a computer. Many programs are available as packages, developed by the various computer-manufacturers, as well as programs developed by specialist organizations. One of the better-known software packages, MILORD, used for calculating resource requirements of MPM networks, provides an example of a computer application. The principle of this program is identical to that applied to the arrow diagram, but the details need to be rigorously specified for the computer, as intuition and visual appraisal are not possible.

There are three phases in this computer program:

(a) The first phase is the calculation of all the activity times and floats, based on the specified resources required for each activity. The calculations of this phase have been described in previous chapters.

(b) The second phase consists of listing in increasing order the latest start-times of activities. If the latest start-times of two or more activities are identical, they are listed in ascending order of total float. Finally, if both these are equal, then the activities can be placed in ascending order of the free float.

(c) The third phase progressively calculates realizable schedules, taking into account the limitations and conditions in the employment of resources. This is done by setting a limit on the resources and rescheduling. The limit is then reset and the cycle is repeated, until either the actual limits are reached or the project duration is extended. The result of this phase is a schedule giving the earliest start-times for each activity and the float available.

Table 6.4 lists the results of the first phase of such a computer analysis, for the house-building project. The resources required have already been specified in Table 3.1. The second stage now lists the activities in ascending order, and the resulting list is shown in Table 8.4.

The activities of obtaining items required for the building do not involve any project resource. Hence, they can be ignored during the calculations involved in resource levelling and/or scheduling; then, for any activity in the list in Table 8.4, there is no preceding activity (in the network) which appears subsequently in the list. Consequently, any change in the scheduled start-time of an activity can affect only the following activities listed.

The first schedule of resource requirements is based on each activity starting at the earliest starting-time. Figure 8.11 shows the histograms for the timing of the requirements of the two resources. On the basis of this schedule, the maximum requirement for craftsmen will be fifteen, over the period days 160–180. The maximum requirement for labourers will be twenty-three over the same period. Appraisal of the activities over this period indicates that activities 16, 17, and 22 have

Table 8.4 *Ordered List of Activities*

Activity	LST	Total float	Free float
10	0		
11	20		
12	40		
13	50		
7	70		
15	80		
6	90		
8	120		
19	160	0	
14	160	80	
1	170		
20	180	0	
18	180	20	
21	210		
23	230		
4	250	250	120
2	250	250	250
5	260		
28	270		
32	290	0	
17	290	130	0
16	290	130	20
3	300		
25	310		
29	320	0	
24	320	110	
27	320	120	
9	320	320	
26	330	130	
22	330	170	
30	340		
33	350	0	
31	350	120	
36	360		
35	380		
37	390		
34	400		
38	410		

substantial floats. From this, the first limits on resources to test are set at seven craftsmen and fourteen labourers.

Referring to the list shown in Table 8.4, activity 17 is the first to be rescheduled, and this is shifted to the first time at

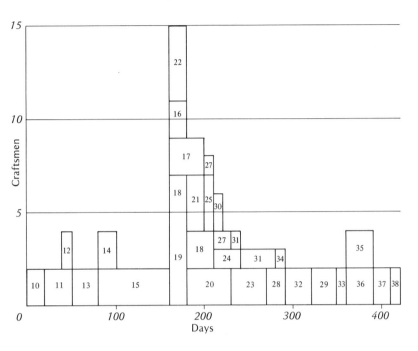

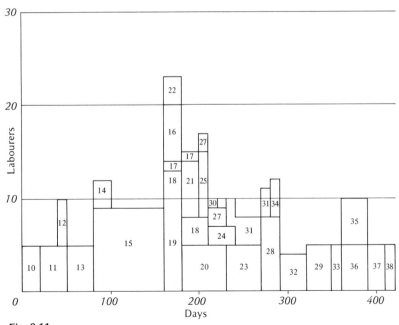

Fig. 8.11

which capacity is available. All activities in the list prior to activity 17 will not be affected by the change in schedule: these are considered to be scheduled at their earliest start-times. Then the first time at which capacity is available will be day 200 (activities 25 and 27 have not yet been scheduled). Once this activity is scheduled, the next activity on the list can be scheduled—this is activity 16, which can begin on day 210. Next, activity 25 is rescheduled, and must occur after activity 16. Capacity is available immediately after completion of activity 16, and so activity 25 can start on day 230. This progress is continued until all the activities after activity 17 (in the list) have been checked—and rescheduled if necessary. One additional step is necessary, if the activity has to follow another, although at that time there are insufficient resources available. In these circumstances, the activity is rescheduled to the next time at which the resources *are* sufficient. Rescheduling activity 30 after activity 17 causes such an overload, and the actual schedule has to be advanced from day 250 to day 260. Table 8.5 gives a list of the rescheduled earliest start-dates in order of calculation. The histograms of the resource requirements are given in Fig. 8.12.

Table 8.5 *Reschedule Calculation: First Cycle*

Activity	Earliest start-date	Activity	Earliest start-date
17	160 to 200	33	350
16	160 to 210	31	230 to 260 (after 22)
25	200 to 260 (after 16)	36	360
29	320	35	360
24	210 to 240	37	390
27	200	34	280 to 310 (after 31)
22	160 to 240	38	410
30	210 to 250 (after 17) to 260		

The cycle of rescheduling is repeated. First, the resources are further limited. From Fig. 8.12, it might appear to be reasonable to attempt to reduce the number of labourers from fourteen to thirteen. With this limit, the first activity requiring rescheduling would be activity 21. Table 8.6, on page 84, lists the rescheduling necessary on shifting activity 21 to the first time at which sufficient resources are available. The histograms of resources are shown in Fig. 8.13.

The last schedule is based on reducing the number of

craftsmen and labourers by one—i.e., limited to six craftsmen, and twelve labourers. Figure 8.14 provides a histogram of the resulting schedule; the total project will be delayed by twenty days if resources are limited that far. Reappraisal with seven craftsmen and twelve labourers, and with six craftsmen and thirteen labourers, will also lead to extension of the duration of

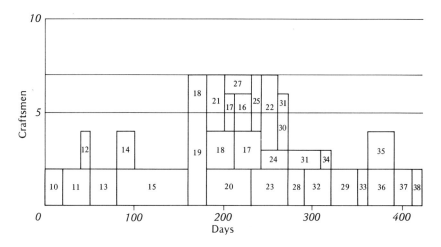

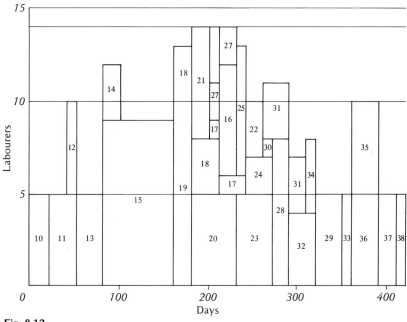

Fig. 8.12

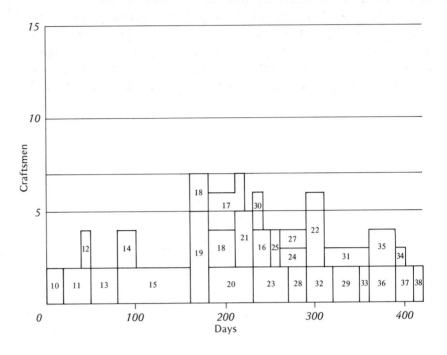
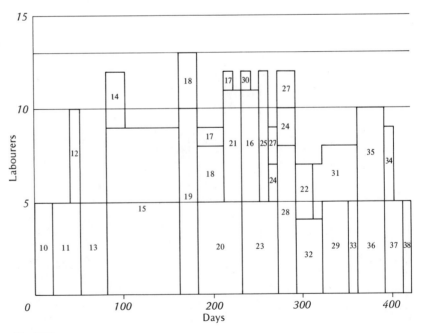

Fig. 8.13

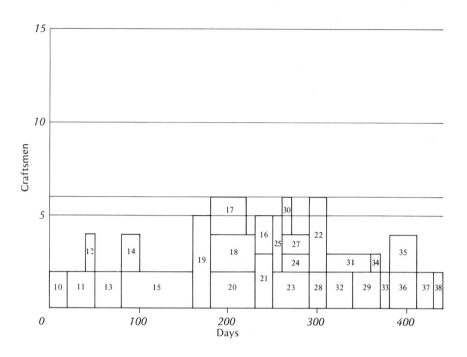
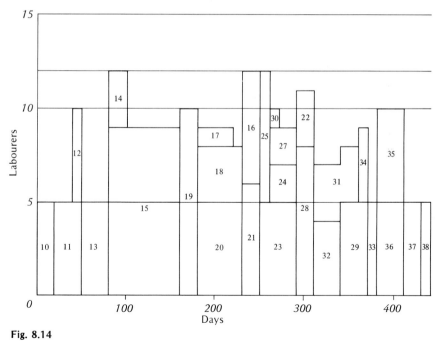

Fig. 8.14

the project. Thus, to complete the project on time, it is necessary to be able to allocate seven craftsmen (compared with fifteen in the initial schedule) and thirteen labourers (compared with twenty-three).

Table 8.6 *Reschedule Calculation: Second Cycle*

Activity	Earliest start-date	Activity	Earliest start-date
21	180 to 210	27	200 to 230 (after 21) to 250
23	230		
28	270	22	160 to 290
32	290	30	210 to 230 (after 21)
17	160 to 180	33	350
16	160 to 230	31	230 to 310 (after 22)
25	200 to 230 (after 21) to 250 (after 16)	36	360
		35	360
29	320	37	390
24	210 to 230 (after 21) to 250	34	280 to 360 (after 31) to 390
		38	410

Checklist

In allocating resources, no optimum solution can be reached by calculation. What method is therefore used to develop the allocations?

Are there any differences between the methods used if: (a) resources are unlimited; and (b) resources are limited? Do you realize how the limiting of resources may affect the project duration?

Is it worth-while considering the use of a computer to determine resource allocations? Why should it be considered? How does a machine like a computer tackle this job?

9
Project control

During the planning stage of a project, the use of network analysis helps to provide information on the action necessary to achieve the objectives. As the network is being developed, decisions on changes in the logic or in the durations will have to be made to bring the planned results in line with the objectives. Then, once the plan has been agreed and the activities started, a constant monitoring of results and updating of the plans must be carried out. Possibly only one thing can be said with certainty about a plan: that it won't actually happen. A project often fails to achieve its objectives because of lack of feedback and updating. Therefore, control must be carefully planned and introduced.

During the execution of a project, actual results must be monitored; and then network analysis will provide up-to-the-minute assessment of the effect of differences between current plans and actual performance. Each time the network is re-analysed, new plans may be produced which will ensure the achievement of the objectives or, at least, evaluate the degree to which the forecast result will deviate from the objectives.

So, at all stages of a project, up to completion, network analysis can provide managers with the necessary information by means of which to control the execution and achieve the objectives.

REVIEW OF NETWORK

Once the draft of the network has been prepared and the schedule of activities defined, the project-team should review all the work done and assumptions made. Even if the project is expected to achieve the objectives set, a check on the validity of the network should be carried out. Checks cover three main areas: the network logic; the estimates of activity durations; and the resource allocations.

Logical Relationships of Activities

The first stage in the review should be the checking of activities and their logical relationships. For each activity, consider:

(a) Whether it is necessary to the project.
(b) Whether, within the network, it can be combined with

others to simplify both planning and control—e.g., install electric plug, and test.

(c) Whether it should be split into two or more sub-activities to provide better control. (This is particularly relevant to activities of long duration on the critical path.)

When all the activities have been reviewed, the logical relationships should be checked, both to ensure that they are correct and that they have been correctly represented in the network diagram.

It may sometimes be difficult to represent accurately the actual planned activities; if so, a compromise has to be accepted. For example, the representation of parallel activities, where overlap is possible, requires some special symbols. However, the possibility of altering the logic to allow overlap of activities might be considered, particularly as regards those on the critical path. Similarly, for critical activities, it may be possible to modify loose logic to shorten the overall duration.

Activity Durations Once the logical representation of activities has been checked, the second stage is to check the estimates of activity durations. One way of doing so is to break down composite activities of long durations into sub-activities. This, in itself, may lead to changes in

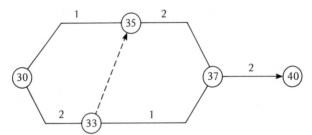

Fig. 9.1

the original estimate. For example, an activity such as installing a processing-unit, estimated to take eight weeks, could be broken down into the sub-network shown in Fig. 9.1. From this network it can be seen that the total duration would be more accurately estimated to take six weeks. The earliest estimate had been based on the manager's assuming a serial approach to the task. Conversely, such a breakdown of activities may show an underestimation of the activity duration (which is generally the case). Another alternative to consider would be changes in design specifications to shorten durations—e.g., prefabrication, or independent sub-assemblies.

The review of activity duration should not be carried out in great detail until the network has been drafted and analysed. When this has been done, activities on the critical path or those with small total float (sub-critical activities) can be reviewed in detail; but those with large float need not be reviewed.

The third stage in the review is to re-appraise the estimated resource requirements, and, if appropriate, the limitations placed on resources. The object is to see if either requirements or limitations can and should be modified. In addition, changes in the technical aspects of the project can be made in order to overcome any constraints placed by resource allocations. Decisions can be made to include extra days over weekends, if possible, and other overtime working. Extra staff may be allocated to tasks, or the work can be sub-contracted.

Resource Requirements

This review is first carried out on the simple, broad network, used at the beginning of the project to assess overall duration and resource requirements. As the review proceeds, so the activities are broken down further and further, developing more accurate estimates of the broad activities' durations. The resource requirements also are more accurately plotted over time, and a better breakdown between different resources is made available. Finally, the project-team will agree that their network provides detail of sufficient accuracy to control the actual execution. The period required for reaching this stage in planning is difficult to evaluate, as it depends on the size of the project and the level of details specified by the team. As an example, it may be possible within two to three weeks to complete a network with 500 activities. Some complex networks may take several weeks between the first, broad plan, drawn on a piece of paper pinned to the office wall, and the final diagram. Even the use of computers at this stage will not speed up the overall duration of planning to any great extent. But the final network diagram, drawn neatly, will show how the project will meet the objectives, and also the timings involved in the work to be carried out.

From Decision to Control

The reason for preparing a network is to aid managers in reaching decisions which should lead to the achievement of project objectives. But this, on its own, is seldom sufficient; it is necessary to exercise control during the actual execution. The fundamental characteristic of a control system is feedback between the operator and the controller—i.e., the comparison of the actual achievement with the plan, and a method of preparing new plans in the light of the current situation.

The first stage of control is, then, to specify a plan of action.

The second is to define the system by which the control information is fed back to the project manager, and the third is to determine the method by which, during execution, modifications to the plans are prepared and implemented. Let us deal with each stage in turn.

SPECIFICATION OF PROJECT PLAN

The plan would be based on the network which has finally been agreed by the team, prior to the actual implementation of activities. Some activities plotted may have already started: for example, preparation of the network diagram. Also, it may have been considered expedient, before the final network was agreed, to initiate some action in order to bring forward the estimated completion date. The final plan of action will include these preliminaries if they are necessary. In essence, however, it will consist of: a list of activities to be completed; a date for starting each activity; and an estimated overall duration.

Schedules of Dates

The scheduled dates of starting activities can be based on their earliest possible start-times and read directly from the network (or printed out directly by a computer). This schedule has the advantage of minimizing the risk that overall project delays may result from certain delays within its execution. To reduce the cost of the project, or at least to extend the timing of cash requirements, it is possible to schedule each activity to start at its latest possible start-time. But such a schedule is inherently risky, because, in the case of activities at the end of sub-networks leading on to critical activities, any delay will extend the overall duration.

Between these two alternatives, any date could be adopted. One further alternative, which provides some measure by which to schedule start-times, is to distribute between all activities any total float within a sub-critical path. But here again, the project manager may utilize some total float unnecessarily during the execution of the project.

A final decision on the date of starting an activity or group of activities has to be made by the project manager. It is probably best if this is done by consideration of the risks associated with such factors as errors in durations estimations, external factors, and resources availability. Thus, if a particular activity requiring no resources (or, possibly, one resource), is on a path with reasonable total float, and if it is almost completely under the control of the project manager, the manager may be justified in scheduling the activity towards the latest start-time. Conversely, activities with very uncertain durations, and/or requiring numerous resources, may be scheduled to start near to the earliest start-time.

In finalizing the schedule, the control structure should be taken into account. It is necessary to decide whether the control should be broadly based or detailed. In some cases, it may be possible to specify only certain critical events and to delegate the detailed control and scheduling to individual managers responsible for that range of activities. If, however, there are substantial inter-relationships between one section and another, it is difficult to delegate a range of activities to one manager. In this circumstance, the project manager will have to specify the schedules in more detail. But whatever system of control is adopted, the managers responsible for implementation of the plan should be consulted, and agreement should be reached with the project manager on the programme of work to be carried out.

Control Structure

Particular events, considered critical to the achievement of the project objectives, might be specifically controlled; in fact, the required result becomes a sub-objective. Such important events occur generally at the start or finish of major stages in execution and are usually called 'milestones'. A black triangle pointing at the appropriate event can be drawn on the network, with a note of the phase to be completed. Figure 9.2 shows such a milestone marked on a network diagram. Within a project, the number of milestones should be kept to a minimum to maintain the value of these specific controls.

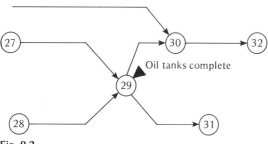

Fig. 9.2

The time horizon of the formal plan has to be decided. There may be a little value in specifying a schedule over a long period, if changes are likely to be made as the earlier activities are executed. Hence, the schedule might provide details of the planned action for three or four activities in sequence ahead, and only broad activity- and event-detail for the balance of the project (using the network of higher levels in the hierarchy of networks).

89

Final Planning in Detail

The form of the final detailed plans would be agreed by the project team and any other managers involved in the implementation. There are many alternative forms:

(a) Activities listed in increasing order of total float.
(b) Activities listed in order of earliest scheduled start-time.
(c) Activities listed by departmental responsibility.
(d) Events in numerical order.
(e) Events by scheduled time.

Details can include network identity, activity number, activity description, estimated duration, scheduled start-time, earliest and latest start-time, earliest and latest end-time, total float, free float, resources required, departments responsible, phase identification.

With projects having a large number of activities, using the arrow and event-network diagram, it is helpful to use event numbers to identify activities. Reference to event numbers, during the execution of the project, then simplifies both the information-flow and the identification of updating required on the diagram. Modifications to the network can also be readily referenced by event numbers, to delete and/or redirect activities. Therefore, starting and finishing event-numbers could be included in the final planning document.

Apart from a summary listing all activities, it is possible to prepare further documentation—e.g., job cards for each activity, and lists (classified by departmental responsibility) of activities to be in progress over the next few periods. It is useful to provide a second copy of the documents used for planning, with space for entering actual achievements. These copies are then used for the project-control information.

PROJECT-CONTROL INFORMATION

In order to control, it is necessary to have information; and information must be timely and its accuracy sufficient for the purposes to which it is put.

What is Needed?

First, it is necessary to specify what control information is needed. This should be as simple as possible. Any events achieved since the last progress report (and the dates of achievement) are the minimum requirement. It is also useful to have an estimate of the earliest and latest time of the end-event of each critical and sub-critical activity which is currently being executed. Sub-critical activities might be defined by the planners as all those with less than a certain total float; for example, if progress reports are to be produced every four weeks, then activities with less than four weeks' total float might be specified as sub-critical.

Secondly, it is important to specify who is to provide the control information. It should be the staff in charge of the actual execution of the activities. Thus, the foreman will report back to the section manager, who reports to the project managers. (There may, of course, be further subdivisions of management responsibility.) The first-level report will probably involve both quantitative and qualitative information, to provide the section manager with a background to the progress against the schedule. The project manager, however, would require details of what action is to be taken to bring any deviations of event times in line with the schedule. If the action is not sufficient, he will have to decide what more must be done. Senior management may then agree to a change in the objectives, or to increasing expenditure to achieve them.

Who Provides It?

A third requirement of a control system is to define the frequency of reporting. For projects lasting a few years, monthly reporting of general progress is probably adequate, though special reports on activities on the critical and sub-critical paths might be needed. In fact, the frequency of reports could vary, but each area of responsibility should have only one frequency to avoid confusion. The frequency could be based on the estimated durations of the activities involved or on the total float available on any sub-critical path, with a maximum frequency for critical activities. At the other extreme, a project lasting a few days might require hourly control information.

How Often?

The frequency of reporting depends on the judgement of the project manager and of the managers involved in implementation. The overall duration of the project, the range of durations of activities, the certainty of the estimates of duration, and the number of critical and sub-critical paths all affect the decision on report frequency. Basically, the control cycle is the judged length of time that non-critical activities can progress without feedback information, irrespective of the actual achievement—i.e., without risk of affecting the achievement of the project objectives. Critical and sub-critical activities can always be reviewed more frequently; but this should be kept to a minimum to avoid over-elaboration of detail at the project-manager level.

Lastly, the timeliness of information through the management structure should be specified in detail. Quite often control information is so out-of-date as to be useless. It must be available in time to change plans and implement them before the next report is due. For example, if the progress reports are prepared monthly, the

Timeliness Necessary

information should be available to the project managers within a few days, and, if necessary, new schedules should be produced within a week. If the reports are prepared daily, information on the progress achieved on each day should be available, already processed, at start of work the following day.

Updating

Once the progress information is available, the network must be updated. If there is little or no difference between the planned and actual results, the network is merely marked with the results to date. Marking can be done physically on the network diagram and/or by updating the information stored within a computer. Even if there has been a delay in non-critical events, the plan might not be changed; but it is necessary to check that subsequent non-critical activities have not become sub-critical.

Any difference between the planned and actual timings of critical or sub-critical activities may involve a re-analysis of the network. If critical events are early, other paths might then be critical. If events are late, some action will be needed to bring the project back in line with the objectives. With delays in sub-critical activities, subsequent activities along the same path will have to be watched more closely to ensure that they do not become critical and affect the completion date.

It can be advantageous to hold a regular 'progress meeting' of the planning staff and the managers involved in the development and implementation of the project. Such a meeting provides for the rapid flow of information on the current situation, allowing immediate development of alternative actions and of decisions made. The frequency of the meetings should be geared to that of the formalized feedback of information on the activity achievements, and the object should be to determine plans for future action.

MODIFICATION DURING EXECUTION

When the information is available and the effect on the network is analysed, the necessary decisions can be made by the project manager. At the regular progress meeting, qualitative assessments of technical aspects can be discussed. Action required and new schedules could be developed and issued at the meeting, which could consider on-going schedules as well, thus bringing the broad analysis of the next period into sharper focus.

The action to be taken might be due to differences between planned and actual results, to modifications in the technical methods specified, or to changes initiated by senior managers. If, at this meeting, it were impossible to reach decisions on any changes of plans necessary, then alternative actions would be evaluated by the planning staff. The results would be discussed at the second meeting, held as soon as possible after the first, and a new plan would then be adopted. If a computer is used, then it may be feasible to evaluate alternatives during the first meeting and reach decisions then.

Instead of holding a second meeting, the project managers may decide on the course of action to be taken, having agreed at the first meeting that any of the alternatives would be acceptable. Then it will be necessary to communicate the new plans to everyone concerned.

The lack of regular re-analysis is one of the major causes of failure of networks to control projects. Within each cycle of information flow, the network may have to be re-analysed twice or more. This leads to a load on the planning-department staff and a time-lag. Computers are ideally suited to re-analysis and can provide the required information quickly. If the project is large, it is recommended that some appropriate computer system be used for the network analysis and project control.

TOTAL PROJECT CONTROL

Figure 9.3 indicates the planning and control cycle in schematic form. To summarize, the cycle starts once the project is sufficiently developed to have resources allocated to it. The first step is for top management to set objectives. The project manager, assisted by senior management and those directly involved, then develops the project plans with the network specialist, using a computer if appropriate. These plans cycle backwards and forwards

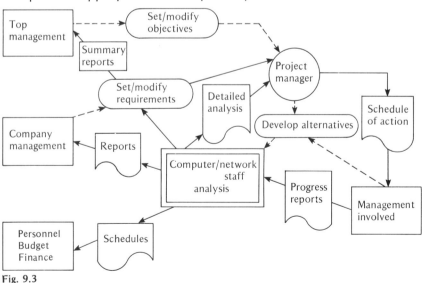

Fig. 9.3

between the project development and the analysis of results. Finally, some summary results are available for top management to review the objectives.

Then, during the implementation of the project, schedules are issued to management involved and to appropriate departments within the organization—e.g., budget and personnel. As results are fed back, so the effects on future plans are analysed and alternative plans are developed if necessary. Further summary information is then passed to top management, and new schedules of work are issued to the managers concerned.

Checklist

The control of a project is a most necessary complement to the planning. Do you have clearly defined control systems within your projects?

The first step in organizing a control system is to have clear and specific plans. Who is responsible for these, and what action does he take to ensure their validity?

Once the plans are formalized, the project will be executed; but the results will not be according to plan. Have you allowed for this in your control system? If so, how are the modified plans prepared?

Do you realize the importance of timely information? Does it have to be absolutely accurate? Who prepared the information? When is it prepared?

All the above questions must be answered before the project starts. Can a computer help with control? Assuming it is used for this purpose, do you appreciate the value of computer analysis?

10
Network analysis within an organization

The theory of network analysis is relatively simple. However, in application, there are a number of problems which face managers within an organization. It is managers who have to provide more information, which must be presented in a rigidly defined and logical form. In return, the technique supplies them with more information for planning and control, and for those managers who are able to make use of this there will be fewer crises, less far-reaching changes in plans, and better fulfilment against objective.

If a project is to be planned and controlled by network analysis, it is necessary to have the assistance of staff experienced in the preparation of detailed networks. Managers involved in the planning and control need to have reasonable knowledge of the diagramming and analysis techiques, and their limitations, while those implementing the project must be able to 'read' a network diagram and the associated schedule-lists. Finally, all management—and particularly top executives—must understand and accept the application of networks to project control. Their continuous support is needed, and especially in the initial stages of introduction. There will be some mistakes and difficulties, which can be used by opponents in efforts to undermine the use of networking, perhaps because they run short of excuses for late completion! However, perseverance will lead to better applications and results, as has been agreed by most companies using the technique on a regular basis. But it is worth repeating that the use of networks is not a panacea. It is no substitute for bad organization, bad management, bad training, or bad information.

TRAINING

The first step in introducing network analysis within an organization is to hold an introductory lecture for senior management, consisting of an appreciation of the outputs obtainable and of the assumptions on which they are based. It is important that senior managers be convinced, at the outset, of the value of network analysis. A half-day course would be sufficient for an initial presentation. A follow-up session may be appropriate, either during or after an actual project, in order to maintain interest and correct misunderstanding. It is important at this stage, and intermittently over the first few projects, to give guidance on the specification and interpretation of summary reports and on the decisions to be made.

For all other managers within a company there should be a two-hour lecture on the basic principles of network analysis, the object being to gain their support and to make them familiar with the terminology, the input requirements, and the outputs. Finally, the implications for management should be clearly stated. Such a lecture would also act as the first step in a more comprehensive training programme for managers directly involved in the development of a project plan. The managers should then participate in a one-day session. This would include some practical exercises as well as more detail of the input requirements and output analysis, showing particularly the effects of changes during the execution of a project.

Finally, a more comprehensive course of, say, three days should be held for potential project managers and for staff directly involved in drawing networks. This is, in fact, all the theoretical training that is needed. The principles of network analysis are very simple to explain; it is in the practice that training is essential. Therefore, the three-day course should include at least one reasonably complete exercise in network diagramming and analysis. A small computer exercise could also be introduced, if it is envisaged that a computer might be used ultimately.

In order to set up and carry out various training lectures and courses, an expert in the application of networking should be involved. In large companies which have taken the decision to implement network analysis extensively, it is possible to employ such a specialist prior to the first project. Generally speaking, however, an outside expert should be engaged for the introductory phase. He would then be able, not only to train internally (and specifically to the needs of the company), but also to aid in initiating and implementing the first project. In addition, he could help with the recruitment of staff skilled in the detailed drawing of networks and, if necessary, in subsequent analysis and preparation of plans.

SELECTING THE FIRST PROJECT The first project should not be the most recent major activity set by the directors for future expansion of the business. That would be jumping in with both feet, without looking! The first project should be reasonably complex, but, as far as possible, self-contained. Once a project is selected, the development of plans should be allowed to progress naturally, and the directors and senior managers should not constantly need to review the situation and visit the project locations; if they do, the staff involved will believe special requirements are involved, and will act to achieve the objectives irrespective of whether there are errors in the network planning.

On completion of the first project, the results should be both measurable and tangible. There is little value in completing a project over a period of time which hardly differs from past or initially expected results. But if it is possible, for example, to complete the publication of a catalogue in one month instead of two, then the technique has got off to a good start.

As soon as feasible after the introduction of network analysis within an organization, a manual of procedures should be drawn up. This manual should specify:

(a) The method of network diagramming and analysis to be used, which includes all terminology and symbols. (The British Standards Institution have brought out a recommended list.)
(b) The forms developed from the network analysis, both in format and in content, together with the types of listing and the control-system information required.
(c) The method of using a computer, if this has been decided upon.
(d) The organization of all staff involved and their responsibilities—this is most important, as it identifies to whom information is given (and its content) and from whom information is required.

USING A COMPUTER

In network analysis, a computer can aid management by quickly producing analysed information in any form required. Indeed, the particular strength of a computer is the extreme rapidity with which it can carry out routine numerical tasks. The analysis of a network, as regards both timings and resource allocations, is just such a task.

If a small project is going to be planned with network analysis, and few re-analyses are likely to be required, there is little value in using a computer. The network of projects with fewer than 400 activities could well be analysed by hand. If there are likely to be several re-analyses, either during the planning stage or during execution (e.g., if the project takes more than one year to complete), a computer should be used for networks of more than 200 activities.

There are many computer programs available for such analyses, and the choice of a particular installation and program depends more on the suitability to a particular company than on the methods by which the calculations are programmed. Typical of the programs are two of those developed by ICL—one to carry out straightforward time analysis, and another to analyse resource allocations. The programs have various features to allow for some of the possible practical requirements—e.g., variable limits on resource availability over a period of time.

Extracting the Data

The preparation of data describing the project for input to a computer is a time-consuming task. And it is advisable, in most applications, to draw the network diagram before extracting the data for computer input. It is easy to make an error, usually of omission, when preparing the logical relationships between activities, and the drawing of networks by hand certainly helps in developing an understanding of the tasks involved and of their interrelations.

Specification of Activities

Although the actual forms required to prepare input data demand careful completion, the contents are similar to the information listed in Fig. 3.1. Apart from titling data of the overall project, each activity would be specified by:

(a) Start- and end-event numbers (some computer programs can do this automatically, if the logical relationship is specified).
(b) Activity number.
(c) Time estimates.
(d) Brief description.

Resource availability/cost card format

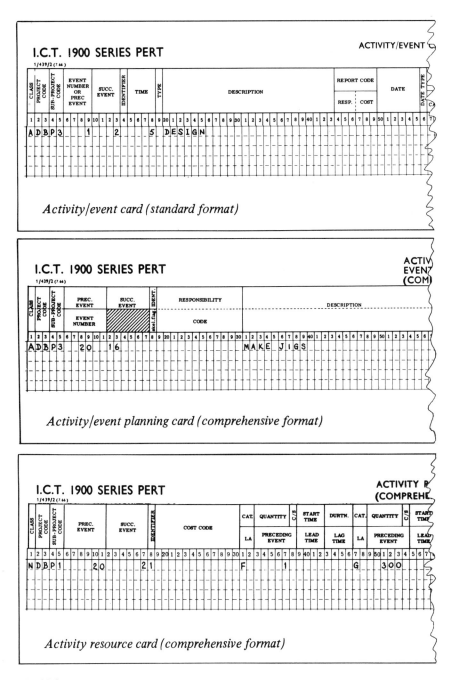

Fig. 10.1

Figure 10.1 shows typical data-sheets which have to be filled in, prior to punching for computer input. Then, if resource-allocation analysis is required, the type of resource and the measure of that resource per period are specified. It may also be necessary to input the resource limitations over the total project.

```
I C L  1900 SERIES PERT                                    21/11/69                    OUTPUT SHEET NUMBER  22

                PROJECT  DB   TEST NETWORK - PST 2.3                          RUN           TIME NOW  5DEC66      PAGE   1

TIME ANALYSIS IN TOTAL FLOAT AND EARLIEST START SEQUENCE
```

S/P CDE	PREC EVENT	SUCC EVENT	REPORT CODE	DESCRIPTION	DUR	EARLIEST START	EARLIEST FINISH	LATEST START	LATEST FINISH	TOT FLOAT	FREE FLT
P3	1	2	TEC	DESIGN	3.0	5DEC66T	22DEC66	2DEC66	21DEC66	-.1	.0
P3	2	4	TEC	PLAN CASTINGS	2.0	22DEC66	9JAN67	21DEC66	6JAN67	-.1	.0
P3	4	14	PUR	OBTAIN CASTINGS	12.0	9JAN67	3APR67	6JAN67	31MAR67	-.1	.0
P3	14	15	PRD	M/C CASTINGS	4.0	3APR67	1MAY67	31MAR67	28APR67	-.1	.0
P3	15	16	QAD	INSPECT	1.0	1MAY67	8MAY67	28APR67	5MAY67	-.1	.0
P3	16	21	PRD	MECH ASSEM	1.0	8MAY67	22MAY67	5MAY67	19MAY67	-.1	.0
P3	21	29	PRD	RUN-IN	1.0	22MAY67	30MAY67	19MAY67	26MAY67	-.1	.0
P3	29	30	QAD	FIT CONTROLS	1.0	30MAY67	6JUN67	26MAY67	5JUN67	-.1	.0
P3	30	31	QAD	TEST	1.0	6JUN67	20JUN67	5JUN67	19JUN67	-.1	.0
P1	1	2	TEC	DESIGN	5.0	5DEC66T	9JAN67	5DEC66T	9JAN67	.0	.0
P1	2	3	TEC	PLANNING	2.0	6JUN67	30JAN67	5DEC66T	9JAN67	.0	.0
P1	13	16	TEC	DESIGN BASE FRAME	10.0	9JAN67	18MAR67	9JAN67	18MAR67	.0	.0
P1	2	13	TEC	DESIGN COVERS	1.0	18MAR67	23MAR67	18MAR67	23MAR67	.0	.0
P1	16	17	PRD	MAKE COVERS M/C1	6.0	23MAR67	8MAY67	23MAR67	8MAY67	.0	.0
P1	17	34	PRD	MAKE CCV M/C 2	6.0	8MAY67	20JUN67	8MAY67	20JUN67	.0	.0
P1	34	23	PRD	FIT COVERS M/C 2	1.0	20JUN67	27JUN67	20JUN67	27JUN67	.0	.0
P1	23	24	QAD	FUNCTION TEST	1.0	27JUN67	4JUL67	27JUN67	4JUL67	.0	.0
P1	24	25	QAD	ACCEPTANCE TEST	2.0	4JUL67	18JUL67	4JUL67	18JUL67	.0	.0
P1	2	6		LEAD	1.0	9JAN67	16JAN67	13JAN67	20JAN67	.4	.0

```
I C L  1900 SERIES PERT                                    22/10/69                    OUTPUT SHEET NUMBER  38

                PROJECT  DB   TEST NETWORK - PST 2.3                          RUN   1       TIME NOW  5DEC66      PAGE   1

RESOURCE-LIMITED ANALYSIS
```

S/P	PREC EVENT	SUCC U EVENT I	REPORT CODE	DESCRIPTION	DUR	EARLIEST START	SCHED START	SCHED FINISH	LATEST FINISH	REM FLOAT	RESOURCES R1	R2
P3	1	2	TEC	DESIGN	3.0	5DEC66	5DEC66T	22DEC66	21DEC66	-.1	4A 3.0	2B 3.0
P1	1	2	TEC	DESIGN	5.0	5DEC66	5DEC66T	9JAN67	9JAN67	.0	3A 5.0	2B 5.0
P3	1	22	TEC	SPECIFY DRIVE MOTOR	1.0	5DEC66	5DEC66T	12DEC66	27JAN67	6.4	1A 1.0	
P3	22	28	PUR	OBTAIN MOTORS	14.0	12DEC66	12DEC66	18MAR67	5MAY67	6.4	*H 13.0 80H 1.0	
P3	1	18	QAD	DESGN & PLN RNNG-IN RIG	2.0	5DEC66	12DEC66	22DEC66	10MAR67	10.4	1A 2.0	1F 2.0
P3	2	4	TEC	PLAN CASTINGS	2.0	22DEC66	22DEC66	9JAN67	6JAN67	-.1	1A 2.0	1B 2.0
P3	2	5	TEC	PLAN COMPONENTS	3.0	22DEC66	22DEC66	16JAN67	17FEB67	4.4	1A 3.0	3B 3.0
P3	22	23	TEC	DESIGN CONTROLS	2.0	12DEC66	22DEC66	9JAN67	17FEB67	5.4	2A 2.0	1B 2.0
P3	18	19 A	PUR	OBTAIN B.O. PARTS	8.0	17DEC66	22DEC66	20FEB67	5MAY67	10.4	*H 7.0 160H 1.0	
P3	2	20	TEC	DESGN ASSEMBLY JIGS	1.0	22DEC66	22DEC66	2JAN67	23MAR67	12.0	1A 1.0	1B 1.0
P3	20	16	PRD	MAKE JIGS	5.4	2JAN67	2JAN67 6FEB67	9JAN67 10MAR67	5MAY67	8.0	1B 5.4 60C 3.0 40C 3.0	1F 5.4

Fig. 10.2

Once the input data are prepared, they can be read into a computer, having the appropriate program, and the outputs required will be obtained. Typical outputs would be as shown in Fig. 10.2. Subsequent changes to the project activities, either during the planning stage or in execution, can be readily referenced by the event numbers of the particular activities involved, and all other activities could be accepted by the computer without change. The re-analysis can then be prepared rapidly.

Bureaux Costs

The costs of using a computer-bureau are very approximately 1p per activity for preparing computer input (but excluding the writing out of the network data in the required form), and, say, £2 per 100 activities for analysis. Thus, a 100-activity network, with one analysis, would cost about £3, while a 500-activity network would cost £16. There is usually a minimum charge of around £5. These costs will, of course, increase over time.

OTHER FACILITIES FOR NETWORKS

The use of networks is continually being expanded into new areas of application, while improving techniques provide more realistic representation of actual activities. Two major areas are concerned: the application of networks to research-and-development projects; and the evaluation of alternative methods of resource allocation. Programs are also being developed to aid in setting up the initial network.

Research and Development

The major problem of drawing a network of research-and-development projects is that the outcome of certain activities may affect the course of those that come subsequently. Therefore, there may be more than one sequence of activities to reach an end-event, and there may be more than one end-event (only one of which will be achieved).

An approach to this problem has been developed by Pritsker under the name of GERT (Graphical Evaluation and Research Technique). In such networks, the alternative paths are specified, and each node is symbolized by an appropriate shape. For alternative (called 'disjunctive') paths, a probability is assigned to each path, depending on the outcome of the previous activity. Then, for each end-event, it is possible to calculate a time, and a probability, that the event will occur.

Resource Allocations

The allocation of resources involves an iterative procedure, which can be very time-consuming. If the range of possible alternative actions is widened, the number of calculations will increase, and the problem will get out of hand. However, there are priority rules which have been shown in practice to provide very good answers. A combination of ordered priority rules might allow consideration of many variables simultaneously.

There are some alternatives which could be tackled in the future. In particular, there is the possibility of splitting activities by one of a number of methods, and also of allocating alternative resources. Such actions would be evaluated on a computer, with the method of priorities specified.

Other Techniques

There are many projects having almost identical sections of activities within the network. In those circumstances, a series of standard sub-networks can be pre-drawn, with only the interlinking activities being specified in detail. If a computer is used, the sub-networks can be readily incorporated to provide a total network, and only the changes to the sub-networks need be specified. Developments of programs to provide networks automatically are available, such as AUTOPERT. Similarly, programs can be used to use computer graph-plotters for drawing networks set up by the computer, or for drawing Gantt charts; AUTONET is one such program.

FINANCIAL CONSIDER-ATIONS

It is difficult to specify the overall cost of applying network analysis to a project, because there are many indirect costs and savings. For example, although the project manager may have to spend some time at the beginning of a project in structuring the activities and their relationships, he is probably only bringing together at one time the effort that would be incurred during the course of a project if he did not use network analysis.

The direct cost of network analysis involves the time taken by draughtsmen to draw the network, and the time taken to analyse the network with the detail required by the project manager. An approximate time-allowance of 20 activities per hour can be used to estimate total draughtsmen-hours required, although it is also necessary to know how many activities are to be specified for the project.

A very approximate figure for the number of activities involved in a project could be based on, say, 3000 to 5000 activities for any large project. Therefore, overall, the manpower to draw up a network for a large project would be 200 man-hours. Once a network is drawn, it is necessary to carry out the evaluation of time—and the sources, if required. It may well involve one week of a senior person's time to analyse the time requirement, but should resource-levelling be required, it would be more economic to use a computer. In addition, during the course of a project, re-analysis of the network needs to be carried out and an overall cost of, say, £2000 per year might be incurred in setting up and running a computer program.

However, the return for the investment in network analysis has always far outweighed the cost. This return is not only based on smoother implementation of the project, but it is of prime importance in the completion of the project more rapidly than would otherwise have been achieved. Thus, the capital is brought into use sooner, and revenue is earned over the period that might otherwise have been absorbed in delays to the project.

To conclude, network analysis does provide better information for the planning and controlling of projects. Accordingly, more logical and prescient decisions are taken by managers, and there is the possibility of substantial improvement in the achievement of project objectives.

Checklist

Does your company or organization prepare managers for the introduction of new techniques? If not, why not?

Have you realized the value of induction of managers to network analysis? A two-hour course would be sufficient for those not directly involved. But, if success is to be achieved, top management must be committed to the application of the technique.

Do you know what contribution an expert can make to the introduction of network analysis within your organization? Can you help in selecting the first project?

Glossary

The British Standards Institution has issued a *Glossary of Terms used in Project Network Analysis* (B.S.4335,1972)

Activity Any task or operation requiring time and possibly other resources; it must be carried out in order to complete a project.

Arrow diagram A diagram representing activities by arrows, and events by circles.

Bar chart See Gantt chart.

Crashing Shortening activity durations by changing the methods specified.

Critical activity An activity which must be completed within the planned duration if the overall project is to finish at the planned time; is on the critical path and has a total float equal to zero.

Critical event An event at the beginning or end of a critical activity; the event-slack is equal to zero.

Critical path A path of critical activities between the start-event and end-event of a project; the critical path determines the overall duration of the project, representing the longest path through the network, and there may be more than one critical path in a network.

Critical-path analysis (CPA) A network-analysis technique, particularly orientated towards determining the minimum project duration.

Critical-path method (CPM) One of the original network-analysis techniques, developed to balance the cost of reducing project duration against savings obtained by so doing.

Dummy activity A logical constraint representing no specific operation and consuming no time; used to indicate relationships between activities.

Duration The estimated, calculated, or actual time required to complete an activity or group of activities.

Earliest event-time The earliest time (or date) at which an event can occur. The preceding activities are assumed to start at their earliest start-times.

Earliest finish-time (EFT) The earliest time (or date) at which an activity can be finished, assuming that it starts at the earliest start-time.

Earliest start-time (EST) The earliest time (or date) at which an activity can start; equivalent to the earliest event-time of the starting-event.

End-event An event with no succeeding activities; it will represent the completion of a project.

Event A point in time marking the start or completion of one or more activities.

Event number A number given to an event for reference purposes, usually such that the preceding event is lower in numerical sequence than the succeeding event. A pair of event numbers can be used for activity identification.

Free float (early) The time by which the start of an activity can be delayed, or by which an activity can be extended, without affecting the earliest event-time of the succeeding event.

Free float (late) The time by which the start of an activity can be delayed, or by which an activity can be extended, without affecting the project duration, even if the preceding event occurs at the latest event-time.

Gantt chart The representation of timings of activities by means of bars drawn against a common time-scale, which can be a relative or an absolute scale.

Independent float The time by which the start of an activity can be delayed, or by which an activity can be extended, without affecting the earliest event-time of the succeeding event, even if the preceding event occurs at the latest event-time.

Lag-time The interval of time between the finish of one activity and the finish of a subsequent activity; used to represent the execution of parallel activities.

Latest event-time The latest time (or date) by which an event must occur, if the project is to finish at the planned time.

Latest finish-time (LFT) The time (or date) by which an activity must finish, if the succeeding event is to occur by the latest event-time; it is equivalent to the latest event-time.

Latest start-time (LST) The time (or date) by which an activity must start, if it is to finish by its latest finish-time.

Lead time The interval of time between the start of one activity and the start of a subsequent activity (irrespective of the activity durations); used to represent the execution of parallel activities.

Metra potential method (MPM) A technique of drawing and analysing a network, in which activities are represented as nodes and logical interrelationships as arrows.

Milestone An event which is of particular relevance within the execution and control of a project. It is usually associated with the completion of a major stage within a project.

Negative float The total float of an activity which is negative. This arises when the activity duration exceeds the time available if the project is to finish on time, particularly during the course of a project when critical activities get delayed.

Network A diagram of arrows and nodes representing activities, events, and their logical interrelationships.

Node A point within a network at which arrows converge and/or from which they diverge.

Parallel activity, ladder activity An activity which can logically start before the completion of the preceding activity and which might be constrained by the completion of that activity.

Preceding event The event at which an activity starts.

Project evaluation and review technique (PERT) An original technique of network analysis, particularly orientated towards control by event-times.

Resource allocation Allocating over time the resources required on a project.

Resource levelling, resource smoothing Scheduling activities within their total float, in order to reduce the variation in resources required over time.

Resource limitation A predetermined limit on the resources available at each period of time within a project duration.

Schedule The specification of the time (or date) at or on which activities are planned to start, the activity durations, and the resources allocated.

Slack, event slack The difference between the earliest and latest event-time.

Start event The event which has no preceding activities; this will represent the start of a project.

Sub-critical path A path through the network activities with small total floats. The total duration of the sub-critical path is not much shorter than the critical path.

Sub-network A section of a network drawn up in greater detail. This is done to provide more information about certain important activities.

Succeeding event An event at which an activity finishes.

Total float The total time by which the start of an activity can be delayed, or by which an activity can be extended, assuming that the preceding event occurs at the earliest event-time and the succeeding event occurs at the latest event-time. This total time is equivalent to the time between the earliest and latest start-time of the activity.

Bibliography

Archibald, R. D., and Villoria, R. L., *Network-Based Management Systems*, John Wiley, New York, 1967.

Armstrong, D. J., et al. *Large Networks*, Operational Research Society, London, 1966. (Report prepared by a working party from the Critical Path Analysis Study Group.)

Battersby, A., *Network Analysis*, (3rd ed.) Macmillan, London, 1970.

Diben, M. L., *Ordonnancement et Potentiels, Methods MPM*, Hermann, Paris, 1970.

Fulkerson, D. R., *Operations Research*, **10**, 1962.

McLaren, K. G., and Buesnal, E. L., *Network Analysis in Project Management*, Cassell, London, 1969.

Moder, J. J., and Phillips, C. R., *Project Management with CPM and PERT*, Reinhold, New York, 1964.

Pritsker, A. A. B., 'GERT Networks', *The Projection Engineer*, October, 1968.

Roy, B., *Révue Française de Récherche Operationelle* **6**, 1962.

Woodgate, H. S., *Planning by Network*, Business Publication Ltd., London, 1967. (The second edition includes network scheduling for production.)

Index

activities, 3, 12, 13, 21-26, 85-87, 98-99
  critical, 45-46
  dummy, 22-25
  parallel, 24-26
  schedule, 36ff, 66ff, 90
  timings, 36ff, 64-65, 88
  uncertainty, 60-61
arrows, 3, 21
arrow diagram, 21ff
AUTONET, 102
AUTOPERT, 102

Bar chart, 4, 64-65

computer, 5, 77, 97-101
construction projects, 8
control;
  event control, 58-60
  project control, 85ff
costs, 102-103
crashing, 86-87
critical events and activities, 45-46
critical path, 46, 51-53, 63
critical path analysis, 2
critical path method, 1, 62-64

dates (for events), 36, 57-58
dummy activities, 22-24
duration, 16-17, 86-87
  uncertainty, 60-61

earliest event times, 36-39
earliest finish times, 41-43, 47-49
earliest start times, 36-37, 41-43, 47-49, 67, 69
errors in diagram, 31

events, 3, 12, 13, 21
  critical, 45-46
  dates, 57-58
  numbering, 26-28
  timing, 36ff, 48

finish times (MPM), 41-43
float, 47-51
  independent, 51
  negative, 58-60
  total, 48, 55, 64, 74, 78
free float (early), 49, 55, 71-72, 78
free float (late), 49

Gautt chart, 4, 64-65, 67, 69, 74-75

histogram (resources), 69ff

independent float, 51
information for control, 90-92
interface, 62

lag-time, 24-26
latest event times, 39-41
latest finish times, 41-43, 47-49, 67
latest start times, 40-43, 47-49
lead time, 24-26
LESS, 2
levelling, 71-74
limited resources, 74-76
logical relationships, 13-14, 85-86

maintenance network, 7
Metra Potential Method (MPM), 3, 31ff
  activity times, 41-43
  floats, 53-55
  resources, 77ff
milestones, 89

MILORD, 77
multi-project analysis, 61-62

negative float, 58-60
net products, services, 9
networks, 3, 20ff
    advantages, 4-5
    control, 89-94
    draughting, 28-29
    hierarchy, 14
    resources analysis, 67ff
    review, 85-88
    time analysis, 36ff
nodes, 21
numbering, 26-28

objectives, 3, 9-10
organisation for networks, 11-12

parallel activities, 24-26
    MPM, 53
PERT, 1, 60
plan for project, 88, 90
precedence diagram, 3

procedures, 97
project management, 11-12, 14, 87-88
    control, 89-94
    training, 95-96
project manager, 11, 89, 92-93

research & development, 8, 101
resources, 16, 87
    allocation, 66ff, 101-102
    levelling, 71-74
    limited, 74-76
    unlimited, 67-71
risk, 1, 60-61

schedule, 88
subcritical activities, 87, 90
subnetworks, 62, 86

time constraints, 24-26, 65
timing, 16-17, 36ff
total float, 48, 55, 64, 74, 78
training, 95-96

uncertainty, 1, 8, 60-61